REEF AND SHORE FAUNA OF HAWAII

CHARLES HOWARD EDMONDSON

REEF AND SHORE FAUNA OF HAWAII

Section 1: Protozoa through Ctenophora

Edited by

DENNIS M. DEVANEY

and

LUCIUS G. ELDREDGE

Bernice P. Bishop Museum Special Publication 64(1)

Bishop Museum Press
Honolulu, Hawaii

PRINTING OF THIS BOOK *was made possible by partial financial support from the Estate of Charles H. Edmondson, to which the Trustees of Bishop Museum wish to express their sincere appreciation.*

Copyright © 1977 by Bernice Pauahi Bishop Museum

All Rights Reserved
Printed in the United States of America
Library of Congress Catalog Card No. 77-89747
ISBN 0-910240-22-1
ISSN 0067-6179

Dr. Frank J. Radovsky, Acting Director
Genevieve A. Highland, Editor
Sadie J. Doyle, Associate Editor

Cover photograph by Richard W. Grigg.
A colony of black coral, Antipathes dichotoma, with a school of Naso hexacanthus, *surgeonfishes known in Hawaii as* "opelu kala."

CHARLES HOWARD EDMONDSON
1876-1970

THIS REVISION of *Reef and Shore Fauna of Hawaii* is dedicated to the memory of Charles Howard Edmondson, the first professionally trained marine zoologist to live in the Hawaiian archipelago. We, the collaborators and editors of the present revision, feel that our use of Dr. Edmondson's title for his handbook on the Hawaiian marine invertebrates is entirely appropriate, for these volumes constitute a continuation of his work, his interests, and his tradition.

C. H. Edmondson was born in Milton, Iowa, on October 14, 1876, and received his undergraduate and graduate training at the University of Iowa, being awarded his doctoral degree from that institution in 1906. In the next seven years he taught at smaller colleges in Iowa, Colorado, and Kansas. He transferred to the University of Oregon in 1913 where he taught until 1920. Through these years, his research work was primarily in temperate freshwater biology.

In 1920 he accepted an appointment as a marine zoologist offered jointly by Bernice P. Bishop Museum and the then young and small College of Hawaii, later to grow into the University of Hawaii, Manoa Campus. The position was shared between the two institutions, since neither had the funds for, or the need of, a full-time marine zoologist. He was also appointed by the College to be the director of the new Cooke Memorial Marine Laboratory, a research and teaching laboratory associated with the Waikiki Aquarium and located on the beach in Kapiolani Park several hundred yards toward town from the present Aquarium. This laboratory, known to the students as the "Beach Lab," was the predecessor and the parent of the several marine institutes and laboratories now operated by the University.

It was through the laboratory in Waikiki that he developed his great research interest in tropical marine biology. His primary fields of interest were the identification of decapod crustaceans and the identification and biology of teredoes or "shipworms." But his interests ranged widely. He studied the ecology of the coral reef off Waikiki, then much more vigorous than now, and the environmental requirements of coral; the control of teredoes and marine fouling organisms; "walking" and stalked jellyfish; regeneration in starfishes; and he even entered the controversy amongst botanists and anthropologists as to whether coconuts had to be carried by man between the islands and archipelagoes of the tropical Pacific, or could drift by currents and grow upon the beach strands. He found that coconuts remained viable for 110 days when floating in seawater, which gave them ample time, he felt, to drift 3,000 miles.

Because Hawaii is the farthest northeastern outpost of the Indo-Pacific marine faunal realm, sweeping from Hawaii across the tropics to the Red Sea, Dr. Edmondson knew that to understand the Hawaiian fauna, he would have to study the animals of the other Pacific Islands. Travel in those days was not by jet aircraft, but by slow ships. He was able to be on two formal expeditions sponsored by the U.S. Navy, the *Tanager,* which visited Johnston and Wake Islands in 1923, and the *Whippoorwill,* which visited the Line Islands in 1924. On one sabbatical leave, he was able to do some marine studies in the Philippines, Singapore, and Ceylon; on another, he visited Fiji and the Society Islands as well as Australia and New Zealand.

He further stimulated marine research by inviting visiting scientists to work at the Waikiki Laboratory—one of them, for example, was C. M. Yonge, the leader of the Great Barrier Reef Expedition of the British Royal Society. He also sent out his collections—he was an avid and discerning field collector—to world experts for study and publication. His graduate students at the University—I was one of them—were stimulated by him to go into marine research.

For twenty-two years, he held his tripartite appointments as Chairman, Department of Zoology, and Director of the Marine Laboratory under the University, and as Marine Zoologist at Bishop Museum. In 1943, having reached the age of compulsory retirement at the University, he devoted full time to his museum work and entered into some of his most productive years, publishing lengthy studies of crabs, marine fouling, and teredoes. He continued his research until 1961 when, at the age of 85, his failing health forced him into complete retirement. In that year, the University of Hawaii awarded to him an honorary degree of Doctor of Science in recognition of his contributions, and in the following year, the new zoology building on campus was named Charles Howard Edmondson Hall.

During his long career in research, he issued more than a hundred publications of considerable scientific merit. However, his *Reef and Shore Fauna of Hawaii* deserves special mention. Early in his stay in Hawaii he must have realized the great need for a semipopular compilation on the inshore fauna for the general populace, and especially for teachers and students. He knew that the general handbooks written for Europe and the United States could not be applied to the tropical fauna of Hawaii and that there were no comparable references written anywhere on the tropical marine biota. He was familiar with the tomes of the specialists in the scientific libraries, but he realized that these were neither available, nor comprehensible, to the interested lay person. He must have devoted some years to the compilation of the first edition, reviewing the scientific literature and the findings of the specialists to whom he had sent his collections; he even had to make his own photographs and drawings. He was met with the problem that many of the Hawaiian species had never been identified. Thus, the only hydroid colony discussed in his second edition could be given only a tentative name and the large jellyfish commonly found in Pearl Harbor could not be named at all. But while the knowledge of the Hawaiian fauna was not adequate to cover all groups uniformly—and it still is not, as this present revision will witness—Edmondson's first (1933) and second (1946) editions were notable for their comprehensive spread and for the fact that almost all of the larger and more conspicuous animals may be identified through the use of his book. His first edition covered only the invertebrates, but his second edition also reviewed the more common fishes. It may be safely said that no one in Hawaii today is capable of such a broad review.

Dr. Edmondson died at the age of 93 in 1970; his wife, Margaret, whom he married in Kansas in 1913, followed him in 1976. They had no children.

The knowledge of the Hawaiian marine fauna, and of the Indo-Pacific marine fauna, is richer by the life work of Charles Howard Edmondson, and, therefore, to his memory the present revision is dedicated.

ALBERT H. BANNER

University of Hawaii
Honolulu, Hawaii
June, 1976

THE CONTRIBUTORS

ALBERT H. BANNER, Hawaii Institute of Marine Biology, University of Hawaii, Coconut Island, P.O. Box 1346, Kaneohe, Hawaii 96744

PATRICIA R. BERGQUIST, University of Auckland, Department of Zoology, Private Bag, Auckland New Zealand

RALPH L. BOWERS, Leeward Community College, Department of Mathematics and Biology, 96045 Ala Ike, Pearl City, Hawaii 96782

WILLIAM JOHNSON COOKE, University of Hawaii, Department of Zoology, Honolulu, Hawaii 96822

CHARLES E. CUTRESS, University of Puerto Rico, Department of Marine Sciences, Mayaguez, Puerto Rico 00708

DENNIS M. DEVANEY, Bernice P. Bishop Museum, P.O. Box 6037, Honolulu, Hawaii 96818

LUCIUS G. ELDREDGE, University of Guam, Marine Laboratory, Box EK, Agana, Guam 96910

RICHARD W. GRIGG, National Sea Grant Program, 3300 Whitehaven St., N.W. Washington, D.C. 20235

E. ALISON KAY, University of Hawaii, Department of General Sciences, Honolulu, Hawaii 96822

JAMES E. MARAGOS, U.S. Army Corps of Engineers, Pacific Ocean Division, Bldg. 230, Fort Shafter, Honolulu, Hawaii, APO San Francisco 96558

DENNIS OPRESKO, 835 N.W. 135th St., Miami, Florida 33168

F.J. PHILLIPS, University of California, Department of Paleontology, Berkeley, California 94720

GERALD E. WALSH, Biological Lab, Bureau of Commercial Fisheries, U.S. Fish and Wildlife Service, Sabine Island, Gulf Breeze, Florida 32561

CONTENTS

Charles Howard Edmondson, 1876–1970 .... Albert H. Banner v
The Contributors viii
Preface .... Dennis M. Devaney xi
Introduction to the Edition of 1946 .... Charles Howard Edmondson 1
Introduction to the Revised Edition .... E. Alison Kay 4

PROTOZOA .... F. J. Phillips 12
   Class Ciliatea 12
      Order Heterotrichida 12
   Class Rhizopodea 15
      Order Foraminifera 15

PORIFERA .... Patricia R. Bergquist 53

CNIDARIA (COELENTERATA) 70
   Class Hydrozoa 70
      Order Hydroida .... William Johnson Cooke 71
   Other Hydrozoans .... L. G. Eldredge and Dennis M. Devaney 105
      Order Siphonophora 105
      Order Chondrophora 105
   Class Scyphozoa .... Dennis M. Devaney and L. G. Eldredge 108
      Order Stauromedusae 109
      Order Cubomedusae 109
      Order Semaeostomeae 111
      Order Rhizostomeae 112
   Class Anthozoa 119
   Subclass Octocorallia .... Dennis M. Devaney 119
      Order Alcyonacea 119
      Order Telestacea 120
      Order Gorgonacea 124
   Subclass Zoantharia 130
      Order Corallimorpharia .... Charles E. Cutress 130
      Order Actiniaria .... Charles E. Cutress 131
      Order Ceriantharia .... Charles E. Cutress 145
      Order Zoanthiniaria .... Gerald E. Walsh and Ralph L. Bowers 148
      Order Scleractinia .... James E. Maragos 158
      Order Antipatharia .... Richard W. Grigg and Dennis Opresko 242

CTENOPHORA .... Dennis M. Devaney 262
    Order Beroida 262
    Order Platyctenea 263
    Order Cydippida 267
    Order Lobata 268
    Order Cestida 268
Index 271

PREFACE

THE FIRST EDITION of *Reef and Shore Fauna of Hawaii* by Charles H. Edmondson, Hawaii's pioneer marine biologist, was published by Bishop Museum in 1933, with a new edition appearing in 1946. Although both editions have been out of print for many years, widespread demand for the book has remained prevalent. Because of the burgeoning interest in marine studies in the intervening years and substantial additions or changes by investigators to the available information about the Hawaiian marine fauna, it was decided that a new and expanded edition was necessary.

About 1963 Dr. Lucius G. Eldredge III, then of Bishop Museum, began discussions with Dr. Edmondson about a new edition of his work. However, Edmondson was then 86 years old and retired and could not prepare the new edition, so he gave Eldredge permission to up-date it for a new edition. Dr. Albert H. Banner of the University of Hawaii was able to obtain a grant from the National Science Foundation (GB3084) in 1965 to permit Eldredge to initiate work on the new edition. It was originally planned merely to bring the names used in the earlier editions up to date and to supplement the text with new records that had appeared in the literature. However, it was soon found that in most groups the changes were too great, and the degree of specialized knowledge required to write the individual sections was beyond the grasp of any one individual. Therefore, experts from the United States, Australia, New Zealand, and the Netherlands were called upon to contribute to the revision. Material from museums, smaller individual collections, specimens received from dedicated students collecting by skin and SCUBA diving and dredging, and information already published had to be utilized. The time and effort contributed by these persons have resulted in more complete data, many new figures and expanded keys, greatly improving this work.

In order to present the underlying concepts that guided Dr. Edmondson in the preparation of the original work, parts of his introduction to the 1946 edition have been included. A new introduction prepared by Dr. E. Alison Kay develops the basis of systematic investigation as an underlying theme of the present revision, and considers the origin of the Hawaiian marine biota, its biogeography, and the varied habitats in which these animals dwell.

In most cases, the bathymetric limits of the fauna in this work extend from the upper levels of the intertidal down to about 100 meters in depth, this being the depth considered as the maximum range of SCUBA. This deeper limit is somewhat arbitrary as far as the fauna is concerned, but does establish a guideline for investigators—below 100 m few reef-building corals occur, the distinctive reef fauna has begun to disappear, as do the living planktonic organisms of the euphotic zone which serve as food for many benthic shallow-water animals. Another realm of island life exists below this level to depths exceeding 3000 m not far from Hawaiian shores and should eventually be the subject of another series of publications.

The format of the revision takes on new approaches in terms of style and content. Instead of a single book, the vast increase in information regarding the Hawaiian and Indo-Pacific marine fauna has required its division into several sections which will be published separately. The various groups of invertebrates are scheduled to appear in the following manner, but not necessarily in the order given below:

Section 1: Protozoa, Porifera, Cnidaria, Ctenophora
Section 2: Platyhelminthes, Nemertina, Aschelminthes,
Ecto-Entoprocta, Brachiopoda, Phoronida
Section 3: Sipuncula, Echiura, Annelida
Section 4: Mollusca
Section 5: Arthropoda
Section 6: Echinodermata, Chaetognatha, Hemichordata,
Chordata (excluding Vertebrates)

All sections are the result of more than one contributor, except for the Mollusca section, which was written in its entirety by Dr. E. Alison Kay.

The revision could not have been completed without the efforts of the contributing authors, who over the intervening years since submitting original manuscripts have requested or have been asked to bring their manuscripts up to date in order to take advantage of new findings, or conform to style changes. We commend them for their patience and dedication. This revision also has benefited from the suggestions and criticisms of a committee composed of Drs. Albert H. Banner, Michael G. Hadfield, E. Alison Kay, S. Arthur Reed, and Sidney J. Townsley, all of the University of Hawaii. Finally, we are indebted to Dr. Eldredge who prepared the groundwork for the project, made initial contact with many of the authors, and organized drafts and suggested the style for the new revision.

Bishop Museum's role in contributing to this project has also been crucial. The Museum's Director (from 1962 to May, 1976), Dr. Roland W. Force, showed an interest and gave encouragement throughout the years of preparation. The present Acting Director, Dr. Frank J. Radovsky, has continued this interest and support, and has been instrumental in bringing this first volume to fruition. The personnel of Bishop Museum Press, Mrs. Genevieve A. Highland, Editor, and Miss Sadie J. Doyle, Associate Editor, devoted many hours to the careful preparation of the manuscript for the printer, and to seeing the book through the press.

It is the hope of everyone concerned that the results of our efforts would have pleased Dr. Edmondson, and that these volumes will serve, for many years to come, the needs of all who are interested in, or who wish to become better acquainted with, the reef and shore fauna of Hawaii.

DENNIS M. DEVANEY

B. P. Bishop Museum
Honolulu, Hawaii
October, 1976

INTRODUCTION
TO THE EDITION OF 1946*
CHARLES HOWARD EDMONDSON

It is not strange to find in an island community such as Hawaii a genuine and widespread interest in the natural history of both land and sea. Our knowledge concerning the inhabitants of the ocean, however, remains relatively meager and fragmentary, because of the obscurity in which they live and the difficulties experienced in gaining information about their habits and modes of life. Although, aside from form and structure, little is known about many common organisms of the sea, by reason of this limitation an inquiry into the intricate relationships of that world of water life becomes the more enticing.

In presenting this condensed pictorial treatise on the marine fauna of Hawaii, the primary purpose is to quicken human interest in the dwellers of the sea, so near at hand, yet so little understood. For many years teachers of biology throughout the territory have been calling for guidance in directing the younger generations in their enthusiastic curiosity about familiar things of their native shores. In an attempt to meet an apparent need, this authoritative information, brief as it may be, about local marine animals was brought together for the benefit of students, teachers, and others who were interested in this particular phase of the organic world in Hawaii.

That serious-minded investigators might know something of the general character, the scarcity or abundance, and the relative accessibility of the main groups of marine animals available for purposes of research about the shores of Hawaii, has also been an important consideration.

Knowledge about Hawaiian marine fauna has been accumulating for more than 110 years, through the efforts of numerous collectors and investigators. Several organized scientific expeditions have made valuable contributions, among them the United States Exploring Expedition in 1840, the *Challenger* Expedition in 1875, and the expedition of the United States Fish Commission, consisting of a shore party in 1901 and the *Albatross* in 1902. To the operations of the *Albatross* is due most of the information known regarding animal life in the waters about Hawaii, from offshore zones down to great depths. The *Tanager* Expedition in 1923 included an extensive exploration of the reefs and shallow waters of the leeward islands of Hawaii as far as Kure (Ocean Island).

Although Hawaii is much nearer the American continent than any other large land area, few shallow-water forms are common to the two regions. The general trend of the shore fauna of the central Pacific area has apparently been from the Indian

---

*Slightly abridged from the original.

Ocean spreading out through the Pacific Ocean to eastern Polynesia, southern Japan, and Hawaii. Some of the common shore forms range all the way from Madagascar and the Red Sea to Hawaii by way of the East Indian route. Hawaii is a northeastern frontier for many species.

Collections of Hawaiian marine fauna are to be found in various museums of Europe and the United States. Probably the most complete is in the United States National Museum, the result of the work of the *Albatross.* There are also valuable collections in Bernice P. Bishop Museum and the University of Hawaii, as well as several extensive privately owned ones, chiefly of Mollusca, in Honolulu.

For the advantage of those who wish to observe marine shore fauna under natural conditions, or to collect them, some of the more accessible and productive localities about the principal islands are mentioned. It is obvious that some species are widely distributed about the shores of all the islands, whereas others are more or less localized. It should also be remembered that many inhabitants of the reefs and shallow water, especially invertebrates, find concealment during the day either in the sand, under stones, or in crevices of rocks. Many forms are, as a general rule, more active at night; if collected during the day one must search for them in their hiding places.

The coast of Oahu is surrounded by a fringing reef with white sand beaches alternating with rocky shores and headlands and indented by numerous bays. Such open reefs as Waikiki and sections of the windward side of the island support a varied fauna. By examining the undersurface of loose stones or breaking up dead coral blocks, one may procure many specimens. An excellent locality for general collecting is the Kahala side of Black Point. Here the sand, the undersurfaces of stones, and the porous rocks yield a great variety of animal forms, among which are echinoderms, worms, crustaceans, and mollusks. In the gravel beds at the south side of Hanauma Bay are to be found species of *Callianassa* [ghost shrimps], annelid worms, gephyreans [sipunculian and echiurian worms], and other burrowing forms; and the loose stones in that vicinity support many other invertebrates. The rough headland of Makapuu and the rocky shores of Waimanalo are interesting to those wishing to collect gastropod mollusks.

Kaneohe Bay harbors a large population of marine animals. Here is one of the best exhibitions of living corals to be seen about the islands, and sponges of many colors are interspersed among coral colonies. The "coral gardens" off Haleiwa, though quite different from those of Kaneohe Bay, rival them in interest. At low tide considerable areas of Kawela Bay are made accessible to the collector. Here is found a varied fauna. On a small reef at Maili Point specimens of the starfish, *Asterope* [=*Asteropsis*] *carinifera,* the holothurian, *Stichopus tropicalis* [=*horrens*], and the brightly colored shrimp, *Stenopus hispidus,* are plentiful.

Pearl Harbor with its extensive shore line is a natural aquarium for many varieties of marine animals. Sponges, clams, and other mollusks, barnacles, and other crustaceans are abundant about the shores, on piling and floats. A large jellyfish is plentiful in the harbor during the winter months. The long snakelike holothurian, *Opheodesoma spectabilis,* is common in Pearl Harbor as well as in Kaneohe Bay.

Many other localities about Oahu are of interest to the marine zoologist. Dredging operations in Honolulu Harbor, at Kewalo, and at Waikiki have yielded forms of animals not usually seen by the shore collector.

Much of the shoreline of Maui is bordered by steep cliffs at the foot of which are lava rocks but no reefs. Such localities support a varied molluscan fauna but are often inaccessible to the collector. The shores of Hana Bay are inhabited by crabs, barnacles, gastropods, and tectibranch mollusks. Along the accessible Maliko coast, on and among the lava stones which line the shore, are to be found gastropod mollusks and crabs. A pulmonate mollusk, *Siphonaria,* with heavy ribs is typical of this region.

One of the best localities for general collecting about Maui is the west side of Maalaea Bay where the sandy beach merges into the rock-covered shore. On the undersurface of the loose stones are many marine organisms. Here in large numbers is a small starfish, *Coscinasterias acutispina,* which undergoes rapid division by fission. Bordering the shore near Makena, Maui, are small reefs accessible at low tide and supporting living corals with the usual fauna typical of such localities. Under the lava stones which line the shore south of Makena lives a small crab, *Cyclograpsus granulatus,* known only from Maui. Here, also, the shore isopod, *Ligia exotica,* is plentiful.

Collecting on Kauai is profitable on the south, east, and north shores. There has been but little investigation along the almost inaccessible Napali coast to the northwest, and long stretches of sandy beaches to the southwest support little animal life. The rocky shores of the Spouting Horn region and those of the east coast in the vicinity of Wailua yield a varied fauna of mollusks, crustaceans, and echinoderms. The north shore from Kalihiwai Bay to Haena is of interest to the collector. On some of the beaches between Hanalei and Haena quantities of marine shells are washed ashore, indicating a rich molluscan fauna in the offshore water. On the shore of Kalihiwai Bay, under the lava stones, are found the isopod, *Ligia kauaiensis,* and several rather uncommon species of shore crabs including *Cyclograpsus henshawi.*

The north, or windward, shore of the long, narrow island of Molokai is quite inaccessible to the collector of marine organisms because of the abrupt cliffs. Along the low southern coast, however, a flat reef platform of considerable width extends the length of the island. Corals grow luxuriantly, especially in the vicinity of Pukoo, and the entire southern shore is of interest to the marine zoologist. Tide pools at the eastern and western extremities of the island are habitats of a varied fauna.

The island of Hawaii is bordered, for the most part, by an exceedingly rough and rugged coast line, making shore collecting hazardous or impossible in many localities. Volcanic action in prehistoric and more recent times is responsible for the present condition of the shore and also for the paucity of reefs about the island. In only a few places and but to a slight extent have corals become established in the near-shore areas. Whereas the more exposed coasts of the island are somewhat barren of animal life except a molluscan fauna characteristic of such an environment, in some protected localities the general collector may find his efforts well repaid. Of numerous places examined on the west coast of Hawaii in 1931, the best collecting was found to be at Honaunau. In this small, sheltered bay bordering the ancient city of refuge is a rich invertebrate fauna. More than 20 species of echinoderms are represented here, and many forms of worms, mollusks, and crustaceans. Localities in the vicinity of Hilo, including the area about Coconut Island and the rocky shore to the southward, are among the most accessible and profitable on the east coast of Hawaii.

INTRODUCTION
TO THE REVISED EDITION

E. ALISON KAY
University of Hawaii

THE MARINE BIOTA of the Hawaiian Islands consists of the descendants of organisms that have been able to disperse successfully over thousands of miles of ocean. Recognized since the mid-nineteenth century as related to that of the Indo-West Pacific—that region of the tropical and subtropical ocean extending from the east coast of Africa in the Indian Ocean to Hawaii, the Line Islands, the Marquesas and Easter Island in the Pacific—the Hawaiian marine biota comprises a distinct component within the Indo-West Pacific, noted especially for its relatively high degree of endemism (Ekman, 1953; Briggs, 1974). The Hawaiian marine biota is also distinguished by attenuation, that is, by having fewer species than are found among other island groups in the western Pacific; by some disharmony, in that the groups of organisms common elsewhere in the region are absent; and by the dominance of some species that are particularly associated with high islands.

Endemism is a characteristic feature of the Hawaiian marine biota and is almost entirely restricted to the species and subspecies levels of the taxonomic hierarchy; there are no endemic families, and only three genera each of mollusks, echinoderms, and fish are thought to be endemic (Kay, in press; Ely, 1942; Gosline and Brock, 1960). At the species level, 18 percent of the algae (Doty, in press); 20 percent of the mollusks (Kay, 1967); 20 percent of the shallow water asteroids and ophiuroids (Ely, 1942); 40 percent of the crustacean family Alpheidae (Banner and Banner, in press); 40 percent of the polychaete worms (Hartman, 1966); and 19 percent to 34 percent of the fish (Randall, in press; Gosline and Brock, 1960) are endemic.

Patterns of endemism are not clear-cut. High shoreline and brackish water mollusks appear to exhibit consistently higher proportions of endemism than do subtidal forms (although 30 percent of the cypraeids and 60 percent of the Turrinae are endemic, Kay, in press). Among the fish, not only are endemics abundant, but "they are represented with surprising regularity among the various reef fish families. There is not a single family represented in Hawaii by ten or more species in which there are no endemics" (Gosline and Brock, 1960). Differences in size and habitat among the mollusks (Kay, 1967) and fin-ray counts among the fish (Strasburg, 1955; Gosline and Brock, 1960), compared with those of the same species elsewhere in their range, suggest incipient speciation or isolation of gene pools. The tiger cowry, *Cypraea tigris,* in Hawaii averages 117 mm in length and is found at depths of more than 10 m; elsewhere in the Pacific the average length of *C. tigris* is 70 mm and the animals are found at depths of a few cm. Among the fish, "when there is a difference in fin-ray counts, the Hawaiian form usually has more fin rays than its representative further south"

(Strasburg, 1955; Gosline and Brock, 1960).

Comparison of the numbers of shallow water species of corals, mollusks, echinoderms, and fish recorded from Hawaii with those found in other island groups within the western Pacific illustrates attenuation. Fifteen genera of corals are recorded from Hawaii, 53 in the Marshall Islands and 40 at Fanning Island in the Line Islands (Maragos, 1972; this volume); 1,000 species of mollusks are found in Hawaii, 2,500 in the Ryukyu Islands (Kay, 1967); 90 echinoderms are known in Hawaii, 345 in the Philippines (Clark and Rowe, 1971); and 450 species of reef and shore fish are recorded from Hawaii, about 1,000 in the Marshall Islands (Randall, pers. comm.).

Several groups of well-known Indo-West Pacific organisms have either never reached the Hawaiian Islands, or were here during the Pleistocene and have since become extinct. *Acropora*, the principal reef-associated coral of the central Pacific is known from Pleistocene raised coral reefs in Hawaii (although recently one small colony of *Acropora* was found at depths of 10 m off the south coast of Kauai (Maragos, this volume)). Among the mollusks, cuttlefish (Sepiidae, Cephalopoda), *Lambis* (Strombidae, Gastropoda), *Tridacna* (Tridacnidae, Pelecypoda) do not now live in Hawaii, while the Vasidae and Haliotidae (Gastropoda) may never have reached Hawaii. Among the fish there are no native snappers of the genus *Lutjanus* or shallow water groupers of the genera *Epinephalus* or *Cephalophis*, genera important in the Indo-West Pacific ichthyofauna elsewhere (Randall and Kanayama, 1973).

The dominance of some marine organisms along Hawaiian shorelines and their absence among the atolls of the central Pacific are also noteworthy. The algae *Ulva* and *Sargassum*, for example, are curiously restricted to high islands (Doty, 1973; Tsuda, 1968). The mollusks *Cellana* (Patellidae, Gastropoda), *Hipponix* (Hipponicidae, Gastropoda), *Littorina pintado* (Littorinidae, Gastropoda), and *Nerita picea* (Neritidae, Gastropoda) have a similar distribution (Kay, in press).

Within an area comprising a chain of islands extending several hundred kilometers from north to south, which encompasses a variety of topographical features and depths, and where average water temperatures differ by as much as 4°C, anomalies in distribution are perhaps not unexpected. Only nine of the 15 genera of Hawaiian reef corals occur in the leeward islands where water temperatures are consistently lower than they are among the windward islands (Ladd, and others, 1967). Fossils dating back to the Miocene of Midway, however, suggest a more diverse coral fauna there than now occurs (Ladd and others, 1967). Several species of endemic mollusks, among them the limpets *Cellana exarata* and *C. sandwicensis* and the muricid *Thais harpa*, are common on basalt shorelines of the windward islands but apparently absent on the calcareous shorelines of the leeward islands (Kay, in press). Conversely, other molluscan species such as *Drupa grossularia* and *Nerita plicata* are common on leeward island shorelines but absent in the windward islands. Among the fish, melanistic forms are common off the Kona coast of Hawaii island but unknown from other areas in the Hawaiian chain (Gosline and Brock, 1960).

HAWAIIAN MARINE ECOSYSTEMS

Benthic marine habitats are traditionally divided into three major zones, the littoral, sublittoral, and the deep sea. The littoral zone has been further subdivided into a littoral fringe, where marine and terrestrial organisms intermingle but where marine organisms predominate, and the eulittoral, occupied by marine organisms adapted to or requiring alternating conditions of submersion and emersion (Lewis, 1964). In the

Hawaiian Islands, where tidal range is limited to less than one meter and where wave action may not only effectively submerge shoreline benches and tide pools but scour sublittoral substrates to depths of more than 15 m, Gosline and Brock (1960) recognize a suprasurge zone, a surge zone, and a subsurge zone in describing the vertical zonation of fish. A modification of Lewis's (1964) scheme and that of Gosline and Brock (1960) is utilized here.

SHORELINE ECOSYSTEMS

The littoral fringe.—The littoral fringe is that area of shoreline fringed by the seaward edge of the maritime vegetation, comprising largely naupaka *(Scaevola),* hau *(Hibiscus),* sea heliotrope *(Messerschmidia),* and *Panicum* in Hawaii. The zone is above the reach of waves and tides but is markedly affected by salt spray. Two regions are distinguishable; an upper region, somewhat localized in occurrence, characterized by broken limestone and/or basalt boulders, and a lower region of more or less continuous rocky substrate of cemented limestone (Emery and Cox, 1956) or basalt. In the upper region where boulders are covered by a canopy of maritime vegetation and the undersurfaces are characterized by conditions of high humidity, at least six species of mollusks (among them the pulmonates *Melampus, Laemodonta,* and *Pedipes)* and the isopod *Ligia* are commonly found. Seaward of the boulder region the shoreline is dominated by two species of littorine, *Littorina pintado* which is widespread in the Indo-West Pacific, and *Nodilittorina picta* which is endemic to Hawaii. Both have a pelagic veliger larval stage during their life histories and hence both are tied to the sea by their mode of reproduction. The littorines are replaced seaward but still above the reach of waves and tides by the black nerite or pipipi, *Nerita picea.* The grapsid crabs, *Pachygrapsus plicatus* and *Grapsus tenuicrustatus,* range through both regions, from the maritime vegetation to the reach of waves and tides.

Basalt shorelines.—Where basalt outcrops extend seaward from the shore, extensive areas of water leveled benches, vertical cliff faces and boulder beaches are prominent features of the coastline on all the windward islands. The shoreward portions of benches and beaches are part of the littoral fringe, but the seaward sections are alternately exposed and immersed by tides twice daily, and scoured by waves seasonally. On basalt benches the highest level of wave action is marked by a line of the crisp red alga, *Ahnfeltia.* Below the *Ahnfeltia* line a variety of frondose algae such as *Ulva, Rhizoclonium, Jania,* and the like cover the substrate. This section of the shore is, in turn, succeeded seaward by a broad band of pink "paint," the growth form of the calcareous alga *Porolithon,* and the interface between the shore and the sea is marked by either the red alga *Pterocladia* or by a mix of the brown seaweeds *Sargassum* and *Ectocarpus.* The dominant mollusks seaward of the *Ahnfeltia* line are the black foot opihi, *Cellana exarata,* the pulmonate limpet *Siphonaria,* the opisthobranch *Smaragdia calyculata,* and the predator gastropods *Morula granulata* and *Thais harpa.* The *Porolithon*-encrusted area of these shorelines is dominated by the yellow foot opihi, *Cellana sandwicensis,* and the shingle urchin, *Colobocentrotus atratus.* The frontal slope is riddled with borings of the sea urchin *Echinometra mathaei,* and a variety of mollusks is found in pockets and crevices of the cliff face: the cowry *Cypraea caputserpentis,* the vermetid *Petaloconchus keenae,* and the bivalve *Isognomon costellatum.* Several carnivorous fish are also associated with this ecosystem, among them the damselfish *Abudefduf imparipennis,* the wrasse *Thalassoma umbrostigum,* and the goby *Bathygobius cotticeps.*

The pattern described above represents the broadest expression of eulittoral zonation found in Hawaii, and it is variously modified on vertical cliff faces, and in sheltered coves and bays. On vertical cliff faces, the *Ahnfeltia* zone and the succeeding frondose algal zone are absent, with the littorines and nerites of the littoral fringe merging directly into the *Porolithon*-encrusted zone. In sheltered coves and bays, especially where there are intrusions of fresh water, the native Hawaiian oyster, *Ostrea sandvicensis*, encrusts vertical substrates between the littoral fringe and the subtidal, and the bivalve *Isognomon californicum* forms dense mats on the horizontal substrates.

Calcareous shorelines.—Calcareous or carbonate shorelines are dominant features of the coastlines of all the major windward islands except Hawaii; on Oahu 52 miles or 32 percent of the coastline is composed of this type of shoreline (Wentworth, 1939).

Topographically solution benches resemble atoll reef flats, consisting of sea level platforms extending from 1 to 30 m seaward from the shore. The benches are separated from shore by a raised, sharply pitted limestone zone and a nip, an indentation at the base of the vertical section. Seaward of the nip the flat-topped surface is densely matted with an algal turf. At the sloping outer edge calcareous algae, and, to a lesser extent, corals, contribute to the structure of the bench. Because of their height above sea level, the surface of the bench may be exposed at low spring tides for periods of as long as four hours.

The biota of calcareous shorelines is distinguished from that of basalt shorelines by its cover of thick algal turf. The turf is interspersed with grazing herbivorous mollusks such as the cowry *Cypraea caputserpentis* and the opisthobranch *Haminoea aperta*, by mats of the suspension feeding mollusks *Brachidontes crebristriatus* and *Dendropoma gregaria*, and by active carnivorous snails such as cones, miters and *Morula*. Both the flora and the fauna are conspicuously zoned. The pools of the pitted zone, which are in effect in the littoral fringe, are inhabited by the small littorine *Peasiella tantilla* and the blenny *Istiblennius zebra*; on the bench itself the cones, especially, form a series from *Conus abbreviatus* at the shoreward edge to *C. chaldaeus* at the seaward edge (Kohn, 1959).

Marine tide pools.—Tide pools occur on sea level basalt outcrops, some formed by depressions in the water-leveled benches, others formed by massive boulders fronting the sea; and on the benches of calcareous shorelines. Physical conditions in marine pools vary with exposure, those pools farthest from the sea undergoing striking variations in temperature and salinity, those at the seaward edge exhibiting essentially subtidal marine conditions. The most exposed pools are characterized by sand substrates bound with blue-green algae. In them are found two or three small mollusks, occasional grapsid crabs, and two fishes, the blenny *Istiblennius zebra* and the goby *Bathygobius fuscus*. Seaward pools are progressively more densely turfed with a variety of algae such as *Padina, Jania,* and *Boodlea*, and there are a variety of worms, mollusks, crustaceans, and echinoderms in them. Seaward pools also serve as incubators for juvenile fishes such as the manini, *Acanthurus sandvicensis*, and the aholehole, *Kuhlia sandvicensis*.

Anchialine pools.—Anchialine pools, shoreline pools without surface connection to the sea but having waters of measurable salinity (0.5 to 30‰) and showing tidal fluctuations, are now found only on the southwest coast of east Maui and the west coast of Hawaii, from Ka'u to Kohala (Maciolek and Brock, 1974). Anchialine pools are recent geological features, subject to obliteration by lava flows and to senescence, the accumulation of organic and mineral deposits originating from aquatic productions and wind-blown materials.

The aquatic vegetation of anchialine pools is dominated by benthic algae such as *Rhizoclonium* or nonencrusting mats of the cyanophytes (blue-green algae) *Scytonema* and *Lyngbya*, as well as the vascular aquatic plant, *Ruppia maritima*. Four decapod crustaceans, two mollusks, and two native fishes are characteristic of the fauna. As on rocky coasts, there is a high degree of endemism: the mollusks, *Theodoxus neglectus* and *T. cariosus*, the small red shrimps *Metabetaeus lohena*, *Procaris hawaiana*, and *Halocaridina rubra*, and the fishes are all endemic to the Hawaiian Islands.

Sandy beaches.—The sand shorelines of the windward Hawaiian Islands are, in general, low, sloping beaches backed by a wall or raised coral platform. Except on Hawaii, sand is largely calcareous, composed of foraminiferan tests, mollusks, echinoderms, and coralline algae (Moberly and others, 1965). On Hawaii there are many black sand and olivine beaches. The most extensive beach development occurs on Kauai; shorter stretches of beach are characteristic of the other islands.

Hawaiian beaches may be subdivided into three zones: an upper beach including the vegetation line; a mid-beach between the high tide line and the vegetation line, its extent dependent on slope and tide; and the lower beach which is continually awash by waves. The biota of sandy beaches is associated with both grain size and slope. The upper beach is characterized by amphipods, isopods, and the males of the ghost crab, *Ocypode laevis*, which burrow in the area (Fellows, 1966). Females of *O. laevis* and males of another species, *O. ceratophthalmus*, burrow in the mid-beach area. The low beach is characterized by the mole crab, *Hippa pacifica*, spionid polychaetes, and four species of the gastropod genus *Terebra* which prey on polychaetes (Miller, 1970). The color of *Hippa, Ocypode,* and *Terebra* is associated with the color of the beach sand: lighter crabs and shells are found on the yellow sand beaches of Oahu, darker forms on the darker beach sands of Maui and Hawaii (Wenner, 1972; Fellows, 1966; Miller, 1970).

Fringing reefs.—Hawaiian reefs are neither so spectacularly developed nor so diverse as are the reefs of other Pacific islands, a circumstance associated with the location of the Hawaiian Islands at the northern edge of the coral reef zone of tropical and subtropical seas, and therefore near the low temperature extreme to which corals are sensitive.

More than half the shoreline of Oahu and comparable portions of Kauai, Molokai, Lanai, and Maui are fringed by reef, but only a small section of the northwest coast of Hawaii, at Kawaihae, is fringed by reef. The reefs are wide, shallow platforms extending as much as 300 meters seaward from the shore. The reef platforms are typically subtidal, one to three meters below mean sea level, although occasional sections may be exposed at low spring tides. The reef flat consists predominantly of sand, coral rubble, and coralline algae. Crustose coralline algae exceed other organisms as the dominant reef builders in Hawaii, with coelenterate corals relatively unimportant in the over-all fringing reef habitat (Littler, 1973). Vigorous coral growth does occur, however, off the frontal edge of the reef flats.

Reef flat assemblages are perhaps the most diverse of those occurring along Hawaiian shorelines, reflecting a variety of habitats: solid substrates of calcareous algae and coelenterate corals, stands of frondose algae such as *Sargassum*, rubble, and sand patches. Because of the variety of habitats, the distribution of reef organisms is patchy: where there are sand patches, infaunal organisms such as the mollusks *Conus pulicarius* and species of *Terebra* occur; where there is rubble or living coral other mollusks, fish, and echinoderms are common. On the Waikiki reef it has been shown that cardinalfish, especially, are associated with different substrates, *Apogon snyderi* feeding over light, sandy substrates, *A. menesmus* over living coral (Chave, 1971).

Estuaries.—Estuaries are defined both as river valleys that receive freshwater discharge, and as the tidal portions of streams. As many as 35 estuarine ecosystems can perhaps be identified in the Hawaiian Islands (Cox and Gordon, 1970). Three major estuaries are Nawiliwili Bay, Kauai, and Pearl Harbor and Kaneohe Bay, Oahu.

Estuarine ecosystems support an endemic fauna of perhaps 38 species, a figure recorded in the estuarine reaches of Kahana Bay, Oahu (Maciolek, pers. comm.). Most estuarine species in Hawaii are euryhaline and most are derived from marine rather than freshwater ancestors (Timbol, 1972). Typical estuarine endemic fishes include the o'opu *(Chonphorous genivittatus)*, o'opu nakea *(Chonophorous stamineus)*, aholehole *(Kuhlia sandvicensis)*, and the mollusks hihiwai *(Theodoxus vespertinus)* and wi *(Neritina granosa)*. Estuaries are also the primary habitats of a few species utilized for food such as the Samoan crab, and they are nursery areas for several inshore marine fishes such as the ama'ama, awa, kaku, and aholehole. Many estuaries in Hawaii are now affected by the invasion of exotic species, for example, the Tahitian prawn, *Macrobrachium lar*, and water hyacinth, *Eichornea crassipea*, which are replacing the native biota.

Mangroves.—Mangroves were introduced on Molokai in 1902 and on Oahu in 1922. On both these islands there are several developed stands which now exhibit many of the properties attributed to mangrove swamps in other tropical areas, but the Hawaiian stands lack the extensive flora and fauna of typical large mangrove stands because of their relatively recent development (Walsh, 1963).

SUBTIDAL ECOSYSTEMS

In addition to coral communities associated with fringing reefs, corals extend subtidally to depths of at least 50 meters in Hawaiian waters, and spectacular coral development occurs on submarine surfaces of recent lava flows off the coasts of the islands of Maui and Hawaii, and in the waters between Maui and Molokai, for example, off Molokini islet and on Penguin Bank. Subtidal coral communities along the Kona coast of Hawaii island appear to be better developed than those off the leeward coasts of others of the windward islands, perhaps because of the large size of Hawaii island and the height of the three major volcanic peaks, Hualalai, Mauna Kea, and Mauna Loa, which provide protection from trade-wind generated seas. Subtidal coral communities are, however, characteristic of all the windward islands, and three distinctive assemblages are recognized.

A *Pocillopora meandrina* assemblage is associated with coastlines where there is considerable wave action and a basalt boulder or rubble substrate, from depths of less than one meter to more than 30 meters. *P. meandrina* is the first coral to appear on new lava flows (Grigg and Maragos, 1974), and this coral forms a dominant element in the shoreline zone off the Kona coast of Hawaii (Dollar, 1973) and in most shoreline areas off Kauai and the windward coast of Oahu. The *P. meandrina* assemblage is often interspersed with other species of corals such as *Porites lobata* and *Montipora verrucosa*, by soft zoanthid corals such as *Palythoa* and *Zoanthus*, and the sea urchins *Echinometra, Echinothrix,* and *Tripneustes*. Fifty-four species of fishes have been reported in the habitat off the Kona coast of Hawaii island (Hobson, 1974).

Coral communities dominated by *Porites lobata* also occur in shallow water. On the Kona coast of Hawaii island this assemblage is found seaward of *Pocillopora meandrina* assemblages; elsewhere it is found in protected bays (Dollar, 1973). *P. lobata* is apparently successful in populating almost any consolidated area from shallow depths down to 30 meters, but modifies its growth form in response to the physical conditions of the environment (Maragos, 1972). Where there is surge, the coral is usually flat and encrusting; in deeper waters the coral occurs as large, lobate, or platelike colonies. Other corals found in the assemblage include *P. meandrina, Fungia scutaria,* and *Porites compressa.* The dominant reef fish in the assemblage are chaetodons, and algae such as *Halimeda* and *Turbinaria* may also be prominent members.

Porites compressa-dominated assemblages are found at depths ranging from less than one meter to more than 50 meters. The colonies of *P. compressa* are distinct from those of *Pocillopora meandrina* and *Porites lobata*, in that the coral is of a branching type, forming fragile thickets that may extend for hundreds of square meters. On the Kona coast of Hawaii island, *P. compressa* assemblages form a zone seaward of *Porites lobata;* elsewhere the corals are prominent in protected bays such as Kaneohe Bay, Oahu, and on Penguin Bank, between Molokai and Maui. Other organisms associated with this assemblage include a variety of reef fish, including chaetodons, and sea urchins such as *Echinometra, Echinothrix,* and *Tripneustes.*

REFERENCES

Banner, A. H., and D. M. Banner
    In press. Alpheidae. In *Reef and Shore Fauna of Hawaii: Section 5, Arthropoda.* Honolulu: Bishop Mus. Press.

Briggs, J. C.
    1974. *Marine Zoogeography.* New York: McGraw-Hill.

Chave, E. H.
    1971. Ecological Requirements of Six Species of Cardinal Fishes (Genus *Apogon*) in Small Geographic Areas in Hawaii. Ph.D. Dissertation, Univ. Hawaii.

Clark, A. M., and F. W. E. Rowe
    1971. *Monograph of Shallow-Water Indo-West Pacific Echinoderms.* London: British Mus. (Natural History).

Cox, D. C., and L. C. Gordon, Jr.
    1970. *Estuarine Pollution in the State of Hawaii.* Vol. 1. *Statewide Study.* Tech. Rep. 31, Water Resources Research Center.

Dollar, S. J.
    1973. Zonation of Reef Corals of the Kona Coast of Hawaii. M. S. Thesis. Univ. Hawaii.

Doty, M. S.
    1973. Marine Organisms: Tropical Algal Ecology and Conservation. In A. B. Costin and R. H. Groves (eds.), *Nature Conservation in the Pacific,* pp. 183-196. Canberra: IUCN and Australian National Univ. Press.
    In press. Algae. In *Enclopedia of Hawaii.* Honolulu: Hawaii Univ. Press.

Ekman, S.
    1953. *Zoogeography of the Sea.* London: Sidgwick and Jackson.

Ely, C. A.
    1942. *Shallow-Water Asteroidea and Ophiuroidea of Hawaii.* B. P. Bishop Mus. Bull. 176. Honolulu.

Emery, K., and D. C. Cox
    1956. Beachrock in the Hawaiian Islands. *Pacific Science* 10: 382-402.

Fellows, D. P.
    1966. Zonation and Burrowing Behavior of the Ghost Crabs *Ocypode ceratophthalmus* (Pallas) and *Ocypode laevis* Dana in Hawaii. M. S. Thesis, Univ. Hawaii.

INTRODUCTION TO THE REVISED EDITION

Gosline, W. A., and V. E. Brock
    1960. *Handbook of Hawaiian Fishes.* Honolulu: Univ. Hawaii Press.
Grigg, R. W., and J. E. Maragos
    1974. Recolonization of Hermatypic Corals on Submerged Lava Flows in Hawaii. *Ecology* 55: 387–395.
Hartman, O.
    1966. Polychaetous Annelids of the Hawaiian Islands. *B. P. Bishop Mus. Occ. Pap.* 23(11): 249–252.
Hobson, E. S.
    1974. Feeding Relationships of Teleostean Fishes on Coral Reefs in Kona, Hawaii. *Fishery Bull.* 72(4): 915–1031.
Kay, E. A.
    1967. The Composition and Relationships of Marine Molluscan Fauna of the Hawaiian Islands. *Venus* 25: 94–104.
    In press. *Reef and Shore Fauna of Hawaii: Section 4, Mollusca.* Honolulu: Bishop Mus. Press.
Kohn, A. J.
    1959. The Ecology of *Conus* in Hawaii. *Ecological Monogr.* 29: 47–90.
Ladd, H. S., J. I. Tracey, Jr., and M. G. Gross
    1967. Drilling on Midway Atoll, Hawaii. *Science* 156 (3778): 1088–1095.
Lewis, J. R.
    1964. *The Ecology of Rocky Shores.* London: English Universities Press.
Littler, M. M.
    1973. The Population and Community Structure of Hawaiian Fringing-Reef Crustose Corallinaceae (Rhodophyta, Cryptonemiales). *J. Experimental Marine Biology* 11: 103–120.
Maciolek, J., and R. E. Brock
    1974. Aquatic Survey of the Kona Coast Ponds, Hawaii Island. UNIHI-Sea Grant-AR-74-04.
Maragos, J. E.
    1972. A Study of the Ecology of Hawaiian Reef Corals. Ph.D. Dissertation, Univ. Hawaii.
    1974. Reef Corals of Fanning Island. *Pacific Science* 28: 247–255.
Miller, B. A.
    1970. Studies on the Biology of Indo-Pacific Terebridae. Ph.D. Dissertation, Univ. New Hampshire.
Moberly, R., Jr., L. D. Baver, Jr., and A. Morrison
    1965. Source and Variation of Hawaiian Littoral Sand. *J. Sedimentary Petrology* 35: 589–598.
Randall, J. E.
    In press. Fishes. In *Encyclopedia of Hawaii.* Honolulu: Hawaii Univ. Press.
Randall, J. E., and R. K. Kanayama
    1973. Marine Organisms: Introduction of Serranid and Lutjanid Fishes from French Polynesia to the Hawaiian Islands. In A. B. Costin and R. H. Groves (eds.), *Nature Conservation in the Pacific,* pp. 197–200. Canberra: IUCN and Australian National Univ. Press.
Strasburg, D.
    1955. North-South Differentiation of Blennid Fishes in the Central Pacific. *Pacific Science* 9: 297–303.
Timbol, A. S.
    1972. Trophic Ecology and Macrofauna of Kahana Estuary, Oahu. Ph.D. Dissertation, Univ. Hawaii.
Tsuda, R.
    1968. Distribution of *Ulva* (Chlorophyta) on Pacific Islands. *Micronesica* 4: 365–368.
Walsh, G. A.
    1963. An Ecological Study of the Heeia Mangrove Swamp. Ph.D. Dissertation, Univ. Hawaii.
Wenner, A.
    1972. Incremental Color Change in an Anomuran Decapod *Hippa pacifica* Dana. *Pacific Science* 26: 346–353.
Wentworth, C. K.
    1939. Marine Bench-Forming Processes: II, Solution Benching. *J. Geomorphology* 2: 3–25.

PROTOZOA

F. J. PHILLIPS*
*Museum of Paleontology,
University of California, Berkeley*

PROTOZOANS are acellular animals, generally microscopic. Many of the functions performed by tissues and organs in more highly evolved animals are performed by subcellular organelles in protozoans. They inhabit fresh and marine waters; some are free-living; many are parasitic; and a number are terrestrial. The free-living forms are well represented in the sea; there are benthonic groups found at most depths and upon most kinds of substrate; and there are planktonic groups which float at various levels. Their study and identification require the use of a microscope. Protozoa fall naturally into four major taxonomic units; the recent classification of Honigberg and others (1964) considers the following units to be subphyla:

Subphylum SARCOMASTIGOPHORA
Subphylum SPOROZOA
Subphylum CNIDOSPORA
Subphylum CILIOPHORA

Other than the Folliculinidae (subphylum Ciliophora) and the Foraminifera (subphylum Sarcomastigophora)—two groups possessing hard shells—the marine protozoans have been poorly studied in Hawaii and are not considered here.

Subphylum CILIOPHORA
Class CILIATEA
Order HETEROTRICHIDA

Family **Folliculinidae**

Eight species of folliculinids are known from Hawaii, but, because the group has been largely overlooked, additional records are to be expected. These "bottle animalcules" are sessile in one stage of life and secrete colored glassy-appearing tubes, or loricae, which are firmly cemented to various substrates. They have been

---

*The author wishes to thank Miss Ruth Todd for reviewing the manuscript and for making valuable suggestions on the species identifications. The majority of the foraminiferal material was collected by URS Research Company of San Mateo, California, for the Hawaiian Electric Company. Additional specimens identified by Dr. G. L. Harrington were provided by Bishop Museum. Stereoscan photographs of foraminifers were taken by Fred Doroshow and Jay Phillips in the Electronics Research Laboratory, University of California, Berkeley, on a machine supported by NIH Grant No. GM-17523. URS Research Company generously provided financial assistance for photography and manuscript preparation. Photographic prints were made by Robert Pitt.

collected from such varied material as submerged wood, the stems and leaves of marine plants, and the external surfaces of living molluscan shells and crustacean carapaces, as well as being established on inorganic substrate. They are known from shallow water estuaries, tide pools, and other marine habitats.

Folliculinid loricae vary in structure from simple to complex. There is often a twofold division of the lorica into sac and neck (Pl. 1). The sac is somewhat variable within a species, depending upon the convolutions of the substrate to which it is attached, but the neck is more consistent in its characters. The morphology of the neck and the presence and nature, if any, of internal valves dividing the lorica are important taxonomic criteria. Color, although not a primary criterion, is often consistent within a species or even a genus, and serves as a useful clue to identification. Loricae often appear in various shades of bottle green in transmitted light and red in reflected light.

Folliculinids have two morphological life stages: a nonfeeding, free-swimming stage, and a sessile, feeding stage with a lorica. A "swimmer," upon settling, usually secretes a new lorica, but specimens have been known to enter an abandoned lorica of the same species. The "swimmer" then metamorphoses into the feeding sessile stage characterized by peristomal lobes on its protoplast. The size, relative proportions, and nature of projections, if any, of the peristomal lobes are taxonomically significant. The shape of the nucleus—either subspherical or moniliform—is also important.

The classification of the folliculinids is based upon the morphology of both the lorica and the protoplast. The swimming stage is difficult to identify, and empty loricae cannot always be identified to species.

Hawaiian records are presented by Andrews (1944) and Matthews (1953, 1962, 1963, 1964) where details of morphology, substrate preference, and general biology are given.

KEY TO HAWAIIAN FOLLICULINIDAE

1. Lorica possesses internal valves 2
   No internal valves 3
2. Lorica differentiated into sac and short oblique neck
   .... *Halofolliculina annulata* (Andrews 1944)
       (Pl. 1, Fig. 4) [see Andrews, 1944; Matthews, 1964 and 1968]
   Lorica rectilinear with a pronounced swelling at base of neck
   .... *Parafolliculina violaceae* (Giard 1888)
       (Pl. 1, Fig. 6) [see Matthews, 1962 and 1968]
3. Lorica not divided into distinct sac and neck; an upright slender tube of
   gradually increasing diameter attached at apex to substrate
   .... *Metafolliculina nordgardi* (Dons 1924)
       (Pl. 1, Figs. 1, 2) [see Matthews, 1962 and 1964]
   Lorica divided into distinct sac and neck 4
4. Spiral whorls on neck of lorica 5
   No spiral whorls on neck of lorica 7
5. Neck oblique to substrate .... *Eufolliculina lignicola* (Faure-Fremiet 1936)
       (Pl. 1, Fig. 7) [see Matthews, 1963 and 1968]
   Neck perpendicular to substrate 6

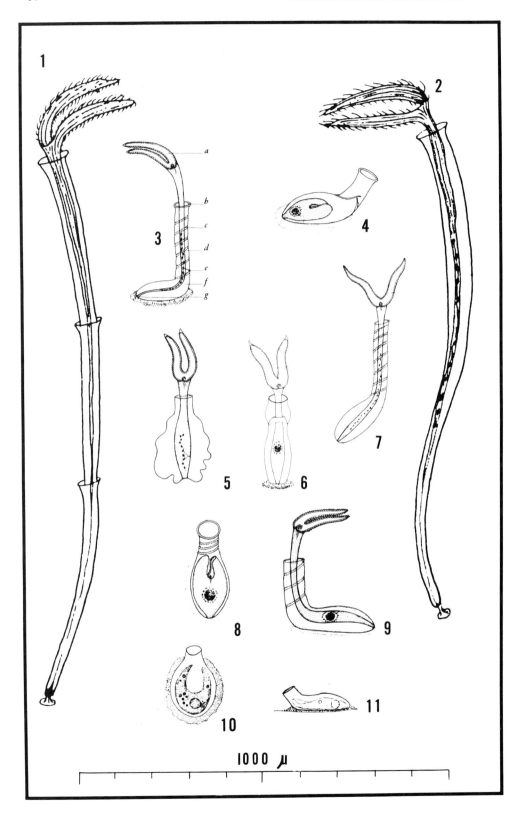

6. Neck short with 2 to 3 spiral whorls .... *Lagotia viridis* Wright 1858
   (Pl. 1, Figs. 8, 9) [see Matthews, 1963, 1964 and 1968]
   Neck long with 6 to 7 spiral whorls ...... *Metafolliculina andrewsi* (Hadzi 1938)
   (Pl. 1, Fig. 3) [see Matthews, 1963 and 1968]
7. Lateral pouches in sac .... *Mirofolliculina limnoriae* (Giard 1883)
   (Pl. 1, Fig. 5) [see Matthews, 1963 and 1968]
   No lateral pouches; neck very short; foot spatulate
   .... *Ascobius simplex* (Dons 1917)
   (Pl. 1, Figs. 10, 11) [see Matthews, 1953]

Subphylum SARCOMASTIGOPHORA
Class RHIZOPODEA
Order FORAMINIFERA

Among the more common Hawaiian marine protozoans are the Foraminifera. They are an order of rhizopods related to the common amoeba, from which they are distinguished by the construction of a test (shell) of one sort or another, by the granular protoplasm, and by the prolongation of the protoplasm into long, thin pseudopodia which branch and anastomose, as opposed to the rather short, stubby pseudopodia of the amoeba. Foraminifera—informally termed "foraminifers," and often shortened to "forams"—are exclusively marine or brackish water organisms that have adapted themselves to nearly every environment of the marine world. Most species are benthonic, either firmly attaching to the substrate or to algae, or possessing free tests and moving slowly about the bottom in search of food. A few species, often occurring in great numbers, are planktonic and float in the open ocean near the surface. Most foraminifers are in the 0.5 mm to 1.0 mm size range, but some are smaller; in tropical regions such as Hawaii, exceptionally large species of 10 mm or greater size occur. For sheer number of individuals, foraminifers are often the most abundant shell-forming organisms found in marine bottom samples.

PLATE 1
PROTOZOA
Family **Folliculinidae**

Figure 1.—*Metafolliculina nordgardi* (Dons). Lateral view of specimen with two extensions to its lorica; after Matthews (1964).

Figure 2.—*Metafolliculina nordgardi* (Dons). Lateral view; after Matthews (1964).

Figure 3.—*Metafolliculina andrewsi* (Hadzi). Lateral view; after Matthews (1968); *a*, peristomal lobe; *b*, collar; *c*, spiral whorl; *d*, neck; *e*, moniliform nucleus; *f*, sac; *g*, foot.

Figure 4.—*Halofolliculina annulata* (Andrews). Lateral view of retracted individual; after Matthews (1968).

Figure 5.—*Mirofolliculina limnoriae* (Giard). Oblique ventral view; after Matthews (1968).

Figure 6.—*Parafolliculina violaceae* (Giard). Lateral view; after Matthews (1968).

Figure 7.—*Eufolliculina lignicola* (Faure-Fremiet). Oblique lateral view; after Matthews (1968).

Figure 8.—*Lagotia viridis* Wright. Oblique ventral view of a contracted specimen; after Matthews (1968).

Figure 9.—*Lagotia viridis* Wright. Lateral view of an extended specimen; after Matthews (1968).

Figure 10.—*Ascobius simplex* (Dons). Ventral view of a contracted specimen; after Andrews (1944).

Figure 11.—*Ascobius simplex* (Dons). Lateral view of the specimen shown in Fig. 10; after Andrews (1944).

All foraminifers possess a test of one sort or another: it may be composed entirely of a pseudochitinous material or may be of some other material laid upon a thin pseudochitinous base. The more complex test wall may be made up of foreign particles selectively chosen by the organism and embedded in a variety of cements, or it may be of calcite secreted by the organism itself. Many of the calcareous tests are perforated by pores distributed over all or a part of the test surface. The test is made up of one, a few, or many chambers, which may be assembled in any of numerous plans of growth. Many tests exhibit different plans of growth in a single specimen. For example, a genus may produce a planospirally coiled test when young but may later add a linear series of chambers to the coiled portion. One or more apertures open into the test from the exterior; these may be of many shapes and are found on various portions of the test. Most of the pseudopodia pass through the aperture. In some calcareous perforate genera, pseudopodia also stream through the pores, but many genera secrete pore plugs which effectively prevent this. The group takes its name from the foramina, or openings, which connect adjacent chambers. These are usually previous apertures but may, in part, result from secondary solution of the chamber wall. Surface ornamentation of almost infinite variety occurs among the various species.

Foraminiferal classification is based upon the nature of the test. Generic and suprageneric taxa are defined in decreasing significance by the possession of various combinations of wall composition, plan of growth, and nature and position of apertures. Foraminiferal species are defined on the basis of ornamentation, test size and shape, and relative proportions of various structural elements.

The empty tests of dead foraminifers are a major constituent of most marine sands in Hawaii. Beach sands are often in large part composed of the broken and tumbled tests of the most robust species, *Amphistegina madagascariensis,* and the thickened central portion of *Heterostegina suborbicularis.* Moberly, Baver, and Morrison (1965, p. 594, and Fig. 3 therein) indicate that in many cases foraminiferal tests compose 50 percent of all littoral sand grains on a given beach. Better preserved tests with a much greater species diversity can be obtained by dredging sand from submerged reef areas or by collecting from a thin surface layer along the strand line. The lightweight tests of foraminifers are sometimes worked into concentrated deposits along the shore by wave action. One of these species, *Marginopora vertebralis,* has received public attention as the "paper or sequin shell" used in the manufacture of one type of shell lei ("Hottest Protozoa Around," *Honolulu Star-Bulletin,* Oct. 14, 1974, p. B-1). Such necklaces have been made for generations by residents of the various Pacific Islands where the species occurs. *M. vertebralis* rarely exceeds 10 mm in diameter in Hawaii, but grows to 25 mm in more tropical areas such as the Great Barrier Reef and Tonga. In Australia it is known as the "mermaid's penny." The construction of these necklaces in Hawaii for sale in the tourist trade forms the basis of a small industry. Occasionally, planktonic species are driven ashore by currents and can be collected in large numbers along the strand line.

FORAMINIFERAL STUDY TECHNIQUES

Dead or preserved foraminiferal tests are best examined dry under a stereoscopic microscope. Magnifications from 20× to 100× are normally sufficient. The dry foraminifer-bearing sediment is sprinkled in a thin layer on a dark tray, to contrast with the predominantly light-colored tests, and individual specimens are manipulated with a moist, fine-pointed artist's brush. Single specimens or assemblages can be glued on 1″ × 3″ cardboard "micropaleontology slides" which are available in a number of formats. The best glue is a thin solution of gum tragacanth (used by druggists as a pill base). It is water soluble, dries clear, and does not shrink and crack specimens upon drying. Glued specimens may be easily reoriented by simply moistening with a wet brush. A drop of formaldehyde in the glue mixture will prevent the growth of bacteria. If muddy samples are to be examined, foraminiferal-sized particles can be concentrated by washing the mud through a 200-mesh-per-inch screen and saving the residue. If foraminifers are rare in a coarse sample, they can be concentrated by floating on heavy liquids, using caution to avoid contact with dangerous vapors. Carbon tetrachloride should not be used as a heavy liquid because it is very dangerous. Alternative heavy liquids include bromoform and perchlorethylene. Care should be taken to keep all equipment clean to avoid contaminating one sample with another.

The traditional technique for the recognition of specimens that were living at the time of collection consists of fixing the sample in 70 percent alcohol, subsequently soaking in a 1-gram-per-liter solution of Rose Bengal dye for 5 minutes, and then washing in water. The dye selectively stains certain proteins red. Care should be taken that a positive stain shows a jelled mass of protoplasm within the test, and not just proteinaceous layers of the test which are preserved long after death. Walker and others (1974) consider this method ineffective and recommend the use of formaldehyde as a preservative and Sudan Black B as a stain. Specimens must not be stored in the formaldehyde preservative, because calcitic components of the test will dissolve with long exposure.

Living specimens that have been collected without physical damage can be examined alive. Many will live for some time in culture dishes, and some have been successfully cultured under laboratory conditions. Some culture techniques are discussed by Arnold (1954, 1974), Ross (1972), and Röttger (1972).

Foraminiferal distribution may be quantified by estimating abundances or by counting specimens. The most commonly used specimen count is the so-called "foraminiferal number," the number of specimens per gram of dry sediment. It is not an ideal statistical tool, however, because it is dependent upon several coincidental variables, such as rate of sedimentation, sediment density, specimen displacement and sorting, and rate of destruction of empty tests, as well as upon such biologically significant factors as rate of reproduction and standing crop. It is largely used by paleontologists who sometimes extend the technique to their neontological studies. A more significant index of abundance is the number of living specimens on or below a unit area of substrate.

Todd (1961) estimates that only 1 percent of the foraminiferal specimens collected in a study of the Gilbert Islands were living when collected; Ross (1972) found that 60 percent were alive in samples from the Great Barrier Reef. The non-living shells represent previous generations that may have been dead for several years. There is a possibility that such empty tests have been displaced by currents from the area in which they were living. A species can be considered to be in place for ecological studies if it is represented by a number of individuals, shows no signs of abrasion or size sorting, and has various growth stages present. A technique has been described for detecting specimens which were living at the time of collection. Live planktonic foraminifers are sometimes obtained in near-shore plankton tows when a fine mesh (less than 0.5 mm) is used, but they are more common in pelagic waters.

HAWAIIAN FORAMINIFERAL FAUNA

While a fairly uniform foraminiferal fauna exists throughout much of the tropical Pacific, Hawaii lies near the northern limit of this tropical region, and some species from the more southerly areas do not reach it. This fauna is designated the Indo-Pacific fauna by Cushman (1948). One distinguishing feature of this area, as well as of other tropical areas, is the presence of large numbers of species in shallow environments. More than 400 foraminiferal species have been recorded from Hawaiian waters less than 100 m in depth and any given sand sample may contain 100 or more species. Similar shallow environments from more temperate regions would contain only a dozen or so species. A second distinguishing feature is the presence of very large species, all of which are restricted to the tropics, and some only to the tropical Pacific. Several of the large endemic foraminifers are relics whose ancestors were widely distributed about the world in the geological past.

Our present knowledge of the Hawaiian foraminiferal fauna is based upon only a few published reports. Although Alcide d'Orbigny mentioned the occurrence of certain species in Hawaii as early as 1826, the first extensive account of Hawaiian forams was that of Brady (1884) (see also Barker, 1960) in his report on the collections of the *Challenger* Expedition. Station 260A of that expedition off Honolulu in 40 fathoms (73 m) remains one of the richest in terms of local reef species, even though many of them had been displaced from their normal habitat into the deeper water where they were collected. Rhumbler (1906) recorded a number of shallow-water species from Laysan Island, and Bagg (1908) recorded deep-water species collected about the islands by the U.S. Bureau of Fisheries vessel *Albatross*. The monumental work of Cushman (1910-1917) summarized these earlier publications and added many additional records from *Albatross* stations and from the numerous samples collected by the *Nero* while sounding the cable line from Honolulu to Midway. Later papers of Cushman (1924, 1925) also record Hawaiian species. Except for Edmondson's (1946) brief mention of a few species, no additional publications have appeared until very recently. Resig (1969, 1974) has studied the marginal marine faunas of Pearl Harbor, Kaneohe Bay, and Salt Lake; she also records Pleistocene fossil Foraminifera (all living species) from wells on the Ewa Plain.

The calcareous imperforate families Miliolidae and Peneroplidae comprise 50 percent of the species found in Hawaii, but most species of these two families—with the exceptions of *Amphisorus hemprichii, Marginopora vertebralis,* and *Sorites marginalis*—occur in small numbers. Where they occur, these latter three species, along with *Amphistegina madagascariensis* and *Heterostegina suborbicularis,* are the dominant species both in number of individuals and in total mass. They are found to some extent in most samples but are particularly abundant in certain areas such as sandy bottoms immediately seaward of coral reefs.

Both benthonic and planktonic foraminiferal species have relatively restricted optimal conditions under which they occur most abundantly; many have wider tolerances, however, and may occur under other conditions in fewer numbers. The primary factors influencing the distribution of benthonic species are depth, temperature, salinity, and nature of substrate. Many genera—particularly those in the families Camerinidae, Peneroplidae, Alveolinellidae, and Amphisteginidae—contain commensal algae (zooxanthellae) within the test. For this reason, these genera are limited to the photic zone and are intolerant of turbidity which clouds the water and blocks light penetration.

In contrast to benthonic genera all species of *Globigerina, Globigerinoides, Globorotalia, Orbulina,* and *Candeina* are planktonic. They are most abundant in a pelagic environment. *Tretomphalus* is planktonic during the later portion of its life cycle; it attaches to algae as a juvenile, however, and is a near-shore species during that stage. The main factors influencing planktonic foram distribution are water temperature, oceanic currents, and water depth. Many planktonic species also contain zooxanthellae. A general treatment of foraminiferal ecology is given by Murray (1973).

Certain species tend to occur together. More general environments can be recognized by the occurrence of these foraminiferal communities, and details can be filled in by considering particular species.

Shallow muddy embayments that may on occasion have reduced salinities are characterized by a distinctive fauna. Resig (1969, Table 4) found that the following species accounted for more than 90 percent of the specimens in five samples from inner Pearl Harbor:

Ammonia beccarii tepida	34%
Quinqueloculina poeyana	30%
Elphidium gunteri galvestonensis	7%
Nonion sp. (given as *Florius*)	7%
Bolivina striatula	7%
Hopkinsina pacifica	3%
Buliminella elegantissima	3%

The species occur in many parts of the world in shallow environments. *Ammonia beccarii tepida* in particular often occurs in large numbers in low-salinity environments where other species cannot survive.

Kaneohe Bay contains most of the above species. Resig (1969, Table 4) found that the following 8 species accounted for 65 percent of individuals in 9 samples from Kaneohe Bay:

Quinqueloculina laevigata
Ammonia beccarii tepida
Nonion sp. (given as *Florius*)
Bolivina striatula
Quinqueloculina bosciana
Quinqueloculina poeyana
Cornuspira planorbis
Discorbis sp.

Because of the greater diversity of habitats in Kaneohe Bay, several other species are present in smaller numbers.

The majority of Hawaiian species are benthonic and occur in shallow water of normal salinity associated with reef habitats. A study in an area off Kahe, on the west coast of Oahu (Phillips, 1973) showed several species being more common in shallower waters and others more common at greater depths.

Species most abundant at Kahe in depths less than 17 m (50 ft) are:

>*Hauerina pacifica*
>*Spiroloculina angulata*
>*Elphidium* spp.
>*Monalysidium politum*
>*Loxostomum limbatum*
>*Reussella* spp.
>*Neoconorbina patelliformis*
>*Rosalina* spp.
>*Ammonia beccarii tepida*
>*Siphoninoides* spp.
>*Cymbaloporetta squammosa*
>*Cibicides* sp.

Species most abundant at Kahe in depths greater than 17 m are:

>*Textularia foliacea oceanica*
>*Quinqueloculina parkeri*
>*Sigmoilina costata*
>*Triloculina bicarinata*
>*Triloculina fichteliana*
>*Operculina philippinensis*

A number of species are most abundant immediately seaward of the reef. It is probable that some of the larger individuals lived on the reef and were displaced by currents. These species are:

>*Heterostegina suborbicularis*
>*Marginopora vertebralis*
>*Spirolina arietina*
>*Amphistegina madagascariensis*

The genera and species keyed and illustrated in the present work are, for the most part, normal reef dwellers found off the west coast of Oahu in the vicinity of Kahe Beach Park from the shore to depths of 35 m. Additional specimens were obtained from Kepuhi Point, Kauai, in 75 m to 150 m on a sand bottom. The BPBM numbers given in the figure captions are those of the Bishop Museum where the figured specimens are deposited. These specimens have been designated hypotypes by the author. The key does not apply to brackish water species such as are found in Pearl Harbor, nor to species that dwell in deep water (greater than about 90 m). For such species the reader is referred to Resig (1969, 1974) and Cushman (1910-1917), respectively.

KEY TO FORAMINIFERAL FAMILIES KNOWN FROM HAWAIIAN WATERS[1]

| 1 | Wall composed of agglutinated particles (Suborder TEXTULARIINA) .... 2 |

---

[1]The familial classification of Cushman (1948) is used with some modifications. Many families not commonly found in shallow Hawaiian waters are not included in the key. Two suborders of foraminifers are not discussed: the Allogromiina, which possess soft tests of pseudochitinous material, and the Fusulinina, an extinct group not found in Hawaii.

	Wall composed of calcareous material, rarely with outer layer of agglutinated material 3
2(1)	Multichambered with biserial plan of growth .... **Textulariidae**
	Multichambered with trochoid plan of growth .... **Trochamminidae**
3(1)	Wall composed of imperforate calcareous material, rarely with outer agglutinated layer (Suborder MILIOLINA) 4
	Wall composed of perforate calcareous material (Suborder ROTALIINA) 7
4(3)	Chambers coiled in various planes; test sometimes with outer agglutinated layer .... **Miliolidae**
	Chambers planispiral, at least in young 5
5(4)	Axis of coiling elongate, tests globular or shaped like rice grains .... **Alveolinellidae**
	Axis of coiling short, test relatively flat 6
6(5)	Chambers usually simple, undivided; later chamber growth assumes many forms ....**Ophthalmidiidae**
	Chambers become annular or uncoiled, often divided into chamberlets .... **Peneroplidae**
7(3)	Aperture usually radiate, test planispiral or uniserial; one genus single-chambered and may have simple round aperture .... **Lagenidae**
	Aperture not radiate, test usually more than one chamber 8
8(7)	Test usually planispiral, at least in young, sometimes becoming trochoid or serial 9
	Test not planispiral in young 12
9(8)	Test consists of proloculus and second coiled undivided chamber ....**Spirillinidae**
	Test multichambered 10
10(9)	Test becoming biserial .... **Heterohelicidae** (part)
	Test coiled throughout or becoming uniserial 11
11(10)	Test involute or partially involute; aperture simple at base of apertural face or cribrate; test sometimes becoming somewhat trochoid .... **Nonionidae**
	Test symmetrically coiled, involute in young but becoming evolute in Hawaiian genera; test generally lenticular, large ...... **Camerinidae**
12(8)	Test serial with aperture loop-shaped or cribrate; rarely single-chambered with slit aperture and internal tube..................... **Buliminidae**
	Test not as above 13
13(12)	Test biserial enrolled; aperture elongate in axis of coiling ..... **Cassidulinidae**
	Test trochoid, at least in young 14
14(13)	Adult dendritic, with irregular branching chambers obscuring early trochoid growth; often red in color .... **Homotremidae**
	Adult not becoming dendritic 15
15(14)	Aperture peripheral or dorsal, at least in young..................... 16
	Aperture ventral, at least in young 17

16(15) Dorsal side often flattened; aperture crosses periphery; growth
remains trochoid in adult .... **Anomalinidae**
Chambers become annular, irregular, or piled upon
one another .... **Planorbulinidae**
17(15) Supplementary chambers between primary chambers on
ventral side.......................................**Amphisteginidae**
No supplementary chambers on ventral side........................ 18
18(17) Chambers inflated, at least in young; test often spinose in life, spines
usually broken off after death; usually coarsely perforate;
aperture large, generally umbilical; planktonic throughout
life cycle 19
Chambers not greatly inflated; not planktonic throughout life cycle...... 20
19(18) Chambers and test as a whole globular; aperture umbilical or along
sutures; spines, if present, elongate and slender; in one form, test
completely enclosed by final spherical chamber .... **Globigerinidae**
Young as above; in adult chambers become flattened on
dorsal side .... **Globorotaliidae**
20(18) Umbilical area open, at least in young.............................. 21
Umbilical area closed or partly or completely filled by plugs 22
21(20) Open umbilicus; trochoid throughout life cycle .... **Discorbidae**
Chambers become annular in adult; generally open umbilicus; final
chamber sometimes covers umbilicus in some
genera .... **Cymbaloporidae**

PLATE 2
PROTOZOA
Family **Textulariidae**

Figure 1.—*Textularia foliacea oceanica* Cushman. BPBM no. A75, *a*, side view; *b*, apertural view; 40X.

Figure 2.—*Textularia siphonifera* Brady. BPBM no. A76; *a*, side view; *b*, apertural view; 40X. This has often been referred to *Gaudryina*, but has no indications of a triserial initial stage.

Figure 3.—*Textularia agglutinans* d'Orbigny. BPBM no. A28; *a*, side view; *b*, apertural view; *c*, edge view; 40X.

Family **Trochamminidae**

Figure 4.—*Trochammina globigeriniformis* (Parker and Jones). BPBM no. A31; *a*, evolute side view; *b*, peripheral view; *c*, involute side view; 100X.

Family **Miliolidae** (in part)

Figure 5.—*Massilina crenata* (Karrer). BPBM no. A77; *a*, side view; *b*, apertural view; 100X. (=*Spiroloculina crenata*).

Figure 6.—*Pseudomassilina* cf. *P. agglutinans* (Keijzer). BPBM no. A78; *a*, side view and *b*, apertural view, 40X; *c*, detail of wall structure, 280X. (=*Massilina agglutinans*).

Figure 7.—*Spiroloculina communis* Cushman and Todd. BPBM A79; *a*, side view; *b*, apertural view; 40X.

Figure 8.—*Spiroloculina angulata* Cushman. BPBM no. A80; *a*, side view; *b*, apertural view; 40X.

Figure 9.—*Spiroloculina corrugata* Cushman and Todd. BPBM no. A81; *a*, side view; *b*, apertural view; 40X.

Figure 10.—*Sigmoilina costata* Schlumberger. BPBM no. A82; *a*, side view; *b*, apertural view; 40X. This species is known from the Atlantic and Mediterranean, but has not previously been reported from the vicinity of Hawaii.

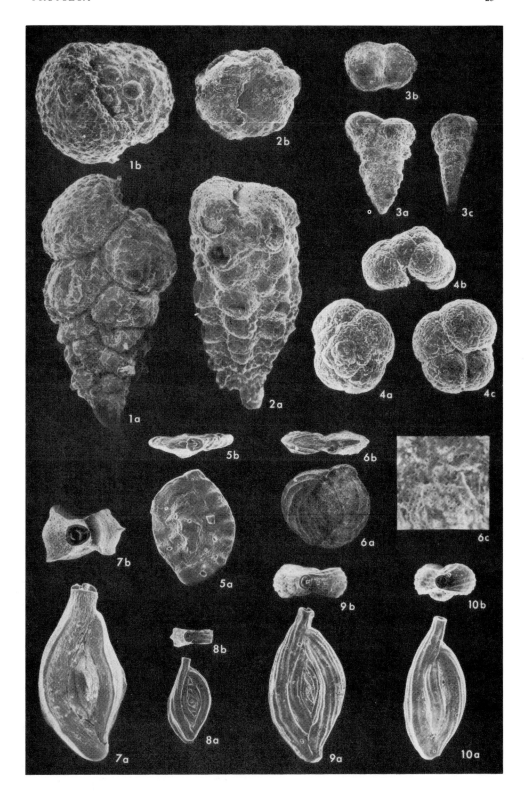

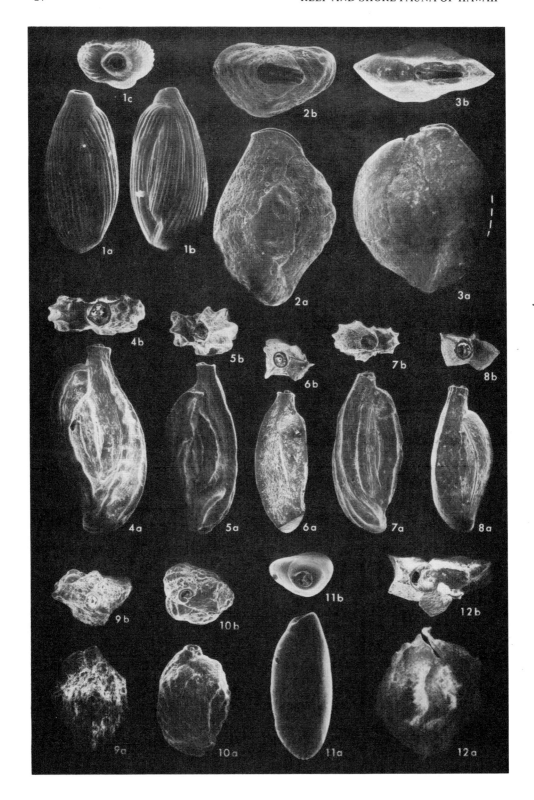

22(20) Test with supplementary calcite on surface; stubby spines not
associated with chamber arrangement .... **Calcarinidae**
Test without supplementary calcite or secondary stubby spines ... **Rotaliidae**

Family **Textulariidae**

This family is characterized by an arenaceous wall and an initial coiled plan of growth which later becomes biserial or, in some genera, uniserial. The only genus of this family commonly found in shallow Hawaiian waters is *Textularia*. The planispiral portion is very much reduced and sometimes is not present. All but the first few chambers are biserially arranged. The aperture is a simple open slit at the inner margin of the last chamber or sometimes in the apertural face [*T. foliacea oceanica* (Pl. 2, Fig. 1); *T. siphonifera* (Pl. 2, Fig. 2); *T. agglutinans* (Pl. 2, Fig. 3)].

Family **Trochamminidae**

An arenaceous wall constructed on a pseudochitinous base and a trochoid plan of growth characterize the family Trochamminidae. The genus *Trochammina* is occasionally found in shallow Hawaiian waters. It has a trochoid plan of growth throughout the life cycle; the aperture is a slit at the inner margin of the involute side of the last-formed chamber [*T. globigeriniformis* (Pl. 2, Fig. 4)].

PLATE 3
PROTOZOA
Family **Miliolidae** (in part)

Figure 1.—*Quinqueloculina poeyana* d'Orbigny. BPBM no. A83; *a* and *b*, opposite side views; *c*, apertural view; 80X.

Figure 2.—*Quinqueloculina parkeri* (Brady). BPBM no. A84; *a*, side view; *b*, apertural view; 40X. (=*Miliolina parkeri*). The marginal crenulations that distinguish the species are not well developed on this specimen.

Figure 3.—*Massilina secans* (d'Orbigny). BPBM no. A85; *a*, side view; *b*, apertural view; 40X. (=*Quinqueloculina secans*).

Figure 4.—*Quinqueloculina ferussacii* d'Orbigny. BPBM no. A86; *a*, side view; *b*, apertural view; 40X.

Figure 5.—*Quinqueloculina sulcata* d'Orbigny. BPBM no. A37; *a*, side view; *b*, apertural view; 40X.

Figure 6.—*Quinqueloculina* sp. BPBM no. A36; *a*, side view; *b*, apertural view; 40X.

Figure 7.—*Quinqueloculina granulocostata* Germeraad. BPBM no. A87; *a*, side view; *b*, apertural view; 40X. The side view shows signs of attack by minute boring organisms.

Figure 8.—*Quinqueloculina polygona* d'Orbigny. BPBM no. A88; *a*, side view; *b*, apertural view; 40X.

Figure 9.—*Quinqueloculina agglutinans* d'Orbigny. BPBM no. A32; *a*, side view; *b*, apertural view; 40X.

Figure 10.—*Triloculina transversestriata* (Brady). BPBM no. A89; *a*, side view; *b*, apertural view; 80X (=*Miliolina transversestriata*).

Figure 11.—*Triloculina* cf. *T. oblonga* (Montagu). BPBM no. A90; *a*, side view; *b*, apertural view; 100X (=*Vermiculum oblongum*, =*Miliolina oblonga*).

Figure 12.—*Quinqueloculina baragwanathi* Parr. BPBM no. A91; *a*, side view; *b*, apertural view; 80X.

Family **Miliolidae**

An imperforate calcareous wall and chambers initially coiled in several planes characterize the Miliolidae. Common Hawaiian genera resemble *Quinqueloculina* in their early growth. All begin with a quinqueloculine plan of growth, but various genera possess different plans of growth as adults.

KEY TO COMMON GENERA OF THE FAMILY MILIOLIDAE

1	Initial quinqueloculine growth stage visible in adult test, sometimes much reduced	2
	Initial quinqueloculine growth stage covered by later chamber growth	10
2(1)	Quinqueloculine growth throughout; simple aperture, usually with tooth .... *Quinqueloculina* [*Q. poeyana* (Pl. 3, Fig. 1); *Q. parkeri* (Pl. 3, Fig. 2); *Q. ferussacii* (Pl. 3, Fig. 4); *Q. sulcata* (Pl. 3, Fig. 5); *Quinqueloculina* sp. (Pl. 3, Fig. 6); *Q. granulocostata* (Pl. 3, Fig. 7); *Q. polygona* (Pl. 3, Fig. 8); *Q. agglutinans* (Pl 3, Fig. 9); *Q. baragwanathi* (Pl. 3, Fig. 12); *Q. bicarinata* (Pl. 5, Fig. 6)]	
	Later growth not quinqueloculine	3
3(2)	Becomes uniserial .... *Articulina* [*A. pacifica* (Pl. 4, Fig. 1)]	
	Does not become uniserial	4
4(3)	Becomes sigmoid, with two chambers per coil added at slightly more than 180 degrees .... *Sigmoilina* [*S. costata* (Pl. 2, Fig. 10)]	
	Becomes planispiral	5
5(4)	Two chambers per planispiral coil	7
	More than two chambers per planispiral coil; aperture cribrate	6

PLATE 4
PROTOZOA
Family **Miliolidae** (in part)

Figure 1.—*Articulina pacifica* Cushman. BPBM no. A92; *a*, side view; *b*, apertural view; 100X. The uniserial plan of growth has not yet developed in this juvenile specimen.

Figure 2.—*Hauerina bradyi* Cushman. BPBM no. A93; *a*, side view; *b*, apertural view; 80X.

Figure 3.—*Hauerina pacifica* Cushman. BPBM no. A94; *a*, side view; *b*, edge view; 80X.

Figure 4.—*Pseudohauerina involuta* (Cushman). BPBM no. A95; *a*, side view; *b*, apertural view; 80X. (=*Hauerina involuta*).

Figure 5.—*Pseudohauerina* sp. BPBM no. A96; *a*, side view; *b*, apertural view; 100X. This is a juvenile of probably either *P. involuta* or *P. orientalis*. The radial pattern formed by the internal partitions is not obvious in this scanning electron micrograph, but is very evident in light microscopic examination.

Figure 6.—*Pyrgo denticulata* (Brady). BPBM no. A35; *a*, side view; *b*, top view; *c*, oblique side view showing aperture; 40X (=*Biloculina ringens* var. *denticulata*).

Figure 7.—*Schlumbergerina alveoliniformis* (Brady). BPBM no. A97; *a*, side view; *b*, apertural view; 80X. (=*Miliolina alveoliniformis*, =*Quinqueloculina alveoliniformis*).

Figure 8.—*Ammomassilina alveoliniformis* (Millett). BPBM no. A98; *a*, side view; *b*, apertural view; 80X (=*Massilina alveoliniformis*). This is a juvenile that has not yet become planispiral.

Figure 9.—*Flintina bradyana* Cushman. BPBM no. A99; *a*, side view; *b*, apertural view; 40X.

6(5) Interior of chambers open .... *Hauerina*
  [*H. bradyi* (Pl. 4, Fig. 2); *H. pacifica* (Pl. 4, Fig. 3)]
  Interior of chambers with transverse partitions .... *Pseudohauerina*
  [*P. involuta* (Pl. 4, Fig. 4); *Pseudohauerina* sp. (Pl. 4, Fig. 5)]

7(5) Aperture simple, with tooth.. 8
  Aperture cribrate; quinqueloculine portion large; arenaceous covering
  on test .... *Ammomassilina*
  [*A. alveoliniformis* (Pl. 4, Fig. 8)]

8(7) Quinqueloculine portion large; planispiral portion
  relatively small.. *Massilina*
  [*M. crenata* (Pl. 2, Fig. 5); *M. secans* (Pl. 3, Fig. 3)]
  Quinqueloculine portion small 9

9(8) Quinqueloculine portion very much reduced or missing; many
  planispiral chambers............................... *Spiroloculina*
  [*S. communis* (Pl. 2, Fig. 7); *S. angulata* (Pl. 2, Fig. 8);
  S. corrugata (Pl. 2, Fig. 9)]
  Quinqueloculine portion obscure; minute pits on surface;
  no true pores .... *Pseudomassilina*
  [*Pseudomassilina* cf. *P. agglutinans* (Pl. 2, Fig. 6)]

PLATE 5
PROTOZOA
Family **Miliolidae** (in part)

Figure 1.— *Triloculina trigonula* (Lamarck). BPBM no. A100; *a*, side view; *b*, apertural view; 80X. (=*Miliola trigonula*.)

Figure 2.— *Triloculina* cf. *T. bicarinata* d'Orbigny of Cushman, Todd, and Post (1954). BPBM no. A101; *a*, side view and *b*, apertural view, 40X; *c*, detail of surface, 500X. Except for the reticulate surface this species is not similar to typical *T. bicarinata*. It is very similar to *Quinqueloculina pseudoreticulata* Parr but it has 3 rather than 5 chambers visible from the exterior.

Figure 3.— *Triloculina fichteliana* d'Orbigny. BPBM no. A102; *a*, side view; *b*, apertural view; 40X.

Figure 4.— *Triloculina linneana* d'Orbigny. BPBM no. A33; *a*, side view; *b*, apertural view; 40X.

Figure 5.— *Triloculina* cf. *T. eburnea* d'Orbigny BPBM no. A56; *a*, side view; *b*, apertural view; 80X. This is similar to *T. oblonga* but the apertural characters are different.

Figure 6.— *Quinqueloculina bicarinata* d'Orbigny. BPBM no. A103; *a*, side view; *b*, apertural view; 100X. This specimen is much smaller than typical *Q. bicarinata* and has more angular carinae.

Figure 7.— *Triloculina oblonga* (Montagu). BPBM no. A104; *a*, side view; *b*, apertural view; 40X. (=*Vermiculum oblongum*, =*Miliolina oblonga*.)

FAMILY **Ophthalmidiidae**

Figure 8.— *Nodophthalmidium antillarum* (Cushman). BPBM no. A105; side view; 40X. (=*Articulina antillarum*.) This species has not previously been recorded from the vicinity of Hawaii.

Figure 9.— *Vertebralina striata* d'Orbigny. BPBM no. A106; *a*, side view; *b*, apertural view; 40X.

FAMILY **Peneroplidae** (in part)

Figure 10.— *Peneroplis planatus* (Fichtel and Moll). BPBM no. A107; *a*, side view; *b*, apertural view; 40X. (=*Nautilus planatus*.) Typical form.

Figure 11.— *Peneroplis planatus* (Fichtel and Moll). BPBM no. A108; side view; 40X. Uncoiling variant.

Figure 12.— *Peneroplis pertusus* (Forskål). BPBM no. A109; *a*, side view; *b*, apertural view; 40X. (=*Nautilus pertusus*.)

10(1)   Triloculine growth plan developed 11
        Five to eight chambers exposed on exterior; aperture cribrate;
            arenaceous covering on test .... *Schlumbergerina*
        [*S. alveoliniformis* (Pl. 4, Fig. 7)]

11(10) Triloculine growth throughout; aperture simple with tooth ...... *Triloculina*
        [*T. transversestriata* (Pl. 3, Fig. 10); *T. trigonula* (Pl. 5, Fig. 1);
        Triloculina cf. *T. bicarinata* (Pl. 5, Fig. 2); *T. fichteliana* (Pl. 5,
        Fig. 3); *T. linneana* (Pl. 5, Fig. 4); *Triloculina* cf. *T. eburnea*
        (Pl. 5, Fig. 5); *T. oblonga* (Pl. 5, Fig. 7); *Triloculina* cf. *T. oblonga*
        (Pl. 3, Fig. 11)]
        Triloculine stage succeeded by fewer chambers per whorl 12

12(11) Planispiral final stage; usually 3 chambers per coil; triloculine stage
            partly visible; aperture simple, large, with tooth .... *Flintina*
        [*F. bradyana* (Pl. 4, Fig. 9)]
        Triloculine stage enveloped by biloculine stage of growth; aperture
            simple with tooth .... *Pyrgo*
        [*P. denticulata* (Pl. 4, Fig. 6)]

Family **Ophthalmidiidae**

The Ophthalmidiidae are distinguished by a calcareous imperforate wall and an initially planospiral plan of growth.

KEY TO COMMON GENERA OF THE FAMILY OPHTHALMIDIIDAE

1   Test consists of 2 chambers, the second being long, tubular, and
        planispirally coiled about the initial chamber .... *Cornuspira*
        [see Pl. 14*a*]
    Test consists of several chambers 2

2(1) Globular initial chamber; planispiral second chamber; final
        chambers uniserial .... *Nodophthalmidium*
        [*N. antillarum* (Pl. 5, Fig. 8)]
    Young trochospiral stage; later uniserial .... *Vertebralina*
        [*V. striata* (Pl. 5, Fig. 9)]

PLATE 6
PROTOZOA
Family **Peneroplidae** (in part)

    Figure 1.—*Amphisorus hemprichii* Ehrenberg. BPBM no. A110; *a*, side view and *b*, edge view, 20X; *c*, detail of edge showing apertures, 100X. Cole considers this species to be the juvenile form of *Marginopora vertebralis*, but others separate the two.
    Figure 2.—*Amphisorus hemprichii* Ehrenberg. BPBM no. A111; oblique side view; 30X. Such malformed specimens are not uncommon. The double layer of chambers and two rows of alternating apertures are visible along the edge.
    Figure 3.—*Marginopora vertebralis* Blainville. BPBM no. A112; *a*, side view and *b*, edge view; 15X; *c*, detail of edge showing apertures, 100X.
    Figure 4.—*Sorites marginalis* (Lamarck). BPBM no. A113; *a*, side view and *b*, edge view, 20X; *c*, detail of edge showing apertures, 100X. (=*Orbitolites marginalis*).
    Figure 5.—*Sorites marginalis* (Lamarck). BPBM no. A114; side view of a broken and regenerated specimen; 20X.
    Figure 6.—*Sorites marginalis* (Lamarck). BPBM no. A115; side view of a juvenile specimen; 100X.

PROTOZOA 31

Family **Peneroplidae**

The Peneroplidae possess an imperforate wall except for the proloculus and second chamber, which are perforate. Early growth is planispiral but later becomes annular or uniserial. The axis of coiling is short, and the tests are relatively flat.

KEY TO COMMON GENERA OF THE FAMILY PENEROPLIDAE

1 Test planispiral or uncoiling in adult; not divided into chamberlets 2
  Test discoid with annular chambers in adult; chambers divided into
    chamberlets 4
2(1) Test generally coiled in adult, relatively flattened and flaring; some
    individuals may tend to uncoil somewhat . . . . *Peneroplis*
    [*P. planatus* (Pl. 5, Figs. 10, 11); *P. pertusus* (Pl. 5, Fig. 12)]
  Test uniserial in adult 3
3(2) Test wall relatively thin, aperture terminal on short neck; coiled portion
    may be absent from common Hawaiian species . . . . *Monalysidium*
    [*M. politum* (Pl. 7, Fig. 3)]
  Test wall relatively thick, no apertural neck . . . . *Spirolina*
    [*S. arietina* (Pl. 7, Fig. 1); *S. acicularis* (Pl. 7, Fig. 2)]

PLATE 7
PROTOZOA
Family **Peneroplidae** (in part)

Figure 1.—*Spirolina arietina* (Batsch). BPBM no. A116; *a,* side view; *b,* apertural view; 40X. (=*Nautilus arietinus,* =*Peneroplis arietinus*).

Figure 2.—*Spirolina acicularis* (Batsch). BPBM no. A117; *a,* side view, 40X; *b,* apertural view, 80X; *c,* detail of side showing pitted surface, 100X. (=*Nautilus acicularis,* =*Peneroplis cylindraceus*).

Figure 3.—*Monalysidium politum* Chapman, BPBM no. A118; *a,* side view, 80X; *b,* apertural view, 250X; *c,* interior of broken first chamber, 250X. Ornamentation consists of longitudinal rows of pits, not tubercules as previously reported by some authors. Figure 3c shows that the pits do not completely penetrate the test wall.

Family **Alveolinellidae**

Figure 4.—*Borelis melo* (Fichtel and Moll). BPBM no. A119; *a,* side view; *b,* end view; 40X. (=*Nautilus melo,* =*Alveolina melo*).

Family **Lagenidae**

Figure 5.—*Vaginulinopsis tasmanica* Parr. BPBM no. A60; *a,* side view, and *b,* apertural view, 40X; *c,* detail of aperture, 100X; *d,* detail of wall structure, 1000X. The terminal chamber has been broken away; 5c shows a cross section of the crystal arrangement in the chamber wall.

Figure 6.—*Lagena globosa* (Montagu). BPBM no. A120; *a,* side view; *b,* apertural view; 200X. (=*Vermiculum globosum,* =*Oolina globosa*).

Family **Camerinidae**

Figure 7.—*Operculina philippinensis* Cushman. BPBM no. A121; *a,* side view; *b,* edge view; 20X. This species has not been previously recorded from the vicinity of Hawaii.

Figure 8.—*Heterostegina suborbicularis* d'Orbigny. BPBM no. A122; *a,* side view; *b,* edge view; 20X.

Figure 9.—*Operculinella cumingii* (Carpenter). BPBM no. A123; *a,* side view; *b,* edge view; 40X. (=*Amphistegina cumingii,* =*Nummulites cumingii*). The last-formed chambers have been broken from this specimen. The flush sutures do not show in this scanning electron micrograph, but are visible under light microscopic examination. There are about 22 chambers per whorl; sutures are limbate, flush, radial near their inner ends, and sharply recurved near the periphery.

PROTOZOA

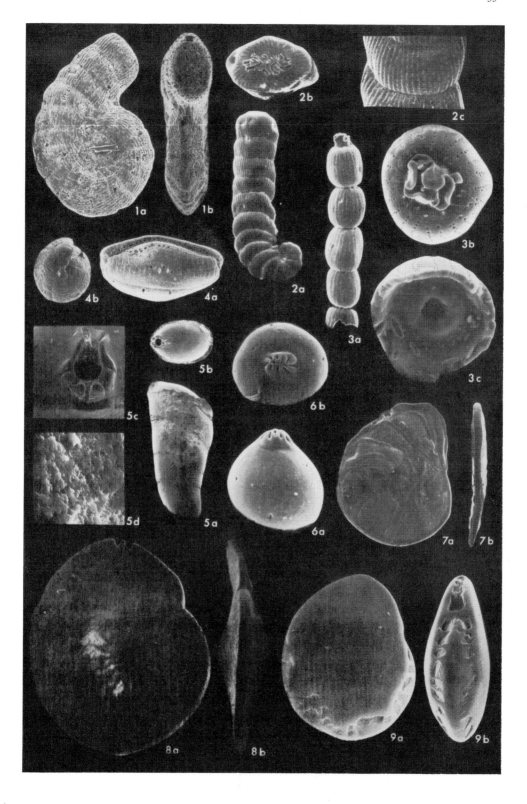

4(1) Chamberlets in single layer; one apertural pore per chamberlet located
around periphery.................................. *Sorites*
[*S. marginalis* (Pl. 6, Figs. 4, 5, 6)]
Chamberlets in more than one layer.............................. 5
5(4) Chamberlets in 2 layers; one apertural pore per chamberlet in 2 alternating
rows around periphery .... *Amphisorus*
[*A. hemprichii* (Pl. 6, Figs. 1, 2)]
Chamberlets in medial layer and 2 lateral layers; numerous apertures
in roughly vertical rows on periphery with 2 horizontal rows, one
at each margin..................................... *Marginopora*
[*M. vertebralis* (Pl. 6, Fig. 3)]

Family **Alveolinellidae**

The Alveolinellidae have a calcareous imperforate wall and are coiled planispirally about an elongate axis. They are either subspherical or rice-grain-shaped (fusiform). Although widely distributed in the geologic record, only two genera are found in the Recent ocean; both live in relatively shallow tropical areas. The genus commonly found associated with shallow reefs in Hawaii is *Borelis* [*B. melo* (Pl. 7, Fig. 4)]. It is subspherical or slightly elongate; it may be distinguished by the single

PLATE 8
PROTOZOA
Family **Nonionidae**

Figure 1.—*Nonion boueanum* (d'Orbigny). BPBM no. A124; *a*, side view; *b*, edge view; 80X. (=*Nonionina boueana*).

Figure 2.— *Pseudononion japonicum* Asano. BPBM no. A125; *a* and *c*, opposite side views; *b*, edge view; 100X.

Figure 3.— *Elphidium* sp. BPBM no. A126; *a*, side view; *b*, edge view; 200X; An immature specimen.

Figure 4.—*Elphidium advenum* (Cushman). BPBM no. A127; *a*, side view; *b*, edge view; 100X. (=*Polystomella advena*).

Figure 5.—*Elphidium poeyanum* (d'Orbigny). BPBM no. A128; *a*, side view; *b*, edge view; 80X. (=*Polystomella poeyana*).

Family **Heterohelicidae**

Figure 6.— *Bolivinella folia* (Parker and Jones). BPBM no. A17; *a*, side view; *b*, apertural view; 80X. (=*Textularia folia*).

Family **Buliminidae** (in part)

Figure 7.—*Buliminella milletti* Cushman. BPBM no. A129; *a*, side view; *b*, apertural view; 150X. The last 5 chambers of the test have been broken away.

Figure 8.— *Bolivina striatula* Cushman. BPBM no. A130; *a*, side view, 100X; *b*, apertural view, 200X.

Figure 9.— *Bolivina compacta* Sidebottom. BPBM no. A42; *a*, side view; *b*, apertural view; 100X.

Figure 10.—*Loxostomum limbatum* (Brady). BPBM no. A131; *a*, side view, 40X; *b*, apertural view, 80X. (=*Bolivina limbata*).

Figure 11.— *Fissurina marginata* (Montagu). BPBM no. A132; *a*, side view; *b*, apertural view; 100X. (=*Vermiculum marginata*, =*Entosolenia marginata*, =*Lagena marginata*). The glossy translucence of the test is not evident in this scanning electron micrograph and the internal entosolenian tube cannot be seen.

Figure 12.— *Uvigerina porrecta* Brady. BPBM no. A133; *a*, side view; *b*, apertural view; 150X.

Figure 13.— *Trifarina* sp. BPBM no. A134; *a* and *c*, side views; *b*, apertural view; 80X.

Figure 14.— *Trifarina bradyi* Cushman. BPBM no. A41; *a* and *c*, side views; *b*, apertural view; 80X.

Figure 15.—*Siphogenerina raphana* (Parker and Jones). BPBM no. A49; *a*, side view, 40X; *b*, apertural view, 100X. (=*Uvigerina raphanus*, =*Sagrina raphanus*, =*Rectobolivina raphana*).

PROTOZOA

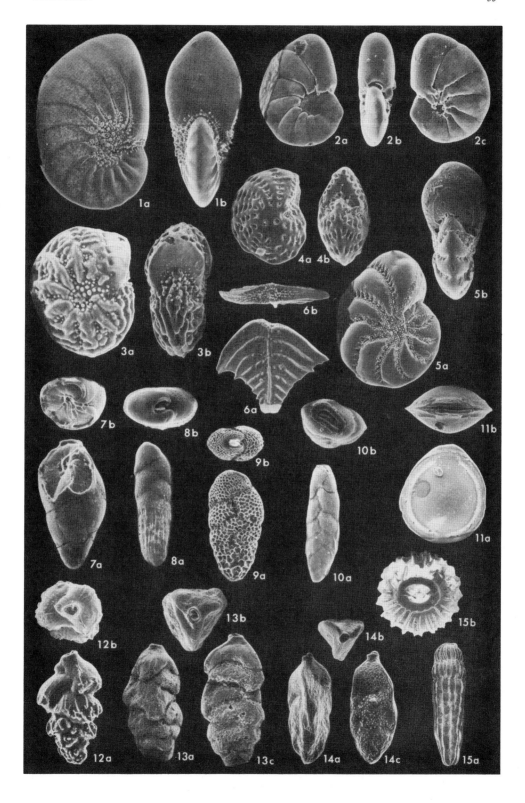

row of apertures. The other Recent genus, *Alveolinella,* is much more elongate and has multiple rows of apertures. It appears to prefer a more tropical environment and does not seem to occur in Hawaii, even though it has been listed by a number of earlier workers.

Family **Lagenidae**

A calcareous perforate test with a radiate aperture, sometimes simple in single-chambered species, and a plan of growth which is either single-chambered, planispiral, or uniserial, but never biserial, distinguishes the family Lagenidae. They are a common constituent of deeper water faunas but are relatively uncommon in shallow water. *Lagena* [*L. globosa* (Pl. 7, Fig. 6)] is consistently found in small numbers in shallow waters. It is single-chambered with a radiate or simple aperture. This common species of *Lagena* is very tiny and globular. *Vaginulinopsis* [*V. tasmanica* (Pl. 7, Fig. 5)] is a deeper water genus. It is planispiral and involute when young but becomes uncoiled as an adult.

Family **Nonionidae**

Most members of the Nonionidae, including those found in Hawaii, are typically planispirally involute with a tendency toward trochoid growth in some genera. The wall is calcareous and perforate. The aperture may be a simple opening or may be cribrate.

KEY TO COMMON GENERA OF THE FAMILY NONIONIDAE

1 Sutures have numerous backward-pointing ridges (retral processes); aperture cribrate or simple............................ *Elphidium* [*Elphidium* sp. (Pl. 8, Fig. 3); *E. advenum* (Pl. 8, Fig. 4); *E. poeyanum* (Pl. 8, Fig. 5)]

PLATE 9
PROTOZOA
Family **Buliminidae** (in part)

Figure 1.—*Reussella simplex* (Cushman). BPBM no. A135; *a,* side view; *b,* apertural view; 80X. (=*Trimosina simplex,* =*Fijiella simplex*). The sutures do not show in this scanning electron micrograph, but closely follow the pattern of the pores.

Figure 2.—*Reussella* cf. *R. aequa* Cushman and McCulloch. BPBM no. A46; *a* and *c,* side views; *b,* apertural view; 80X.

Figure 3.—*Reussella spinulosa* (Reuss). BPBM no. A136; *a,* side view; *b,* apertural view; 80X. (=*Verneuilina spinulosa*).

Figure 4.—*Chrysalidinella dimorpha* (Brady). BPBM no. A18; *a,* side view; *b,* apertural view; 80X. (=*Chrysalidina dimorpha*).

Figure 5.—*Pavonina flabelliformis* d'Orbigny. BPBM no. A137; *a,* side view; *b,* apertural view; 40X; an adult specimen.

Figure 6.—*Pavonina flabelliformis* d'Orbigny. BPBM no. A22; *a,* side view; *b,* apertural view; 80X; a juvenile.

Family **Spirillinidae**

Figure 7.—*Planispirillina denticulogranulata* (Chapman). BPBM no. A138; *a* and *c,* opposite side views; *b,* edge view; 80X. (=*Spirillina denticulo-granulata*).

Family **Discorbidae** (in part)

Figure 8.—*Rosalina* cf. *R. vilardeboana* d'Orbigny. BPBM no. A139; *a,* evolute side view; *b,* edge view; *c,* involute side view; 100X.

Figure 9.—*Rosalina orientalis* (Cushman). BPBM no. A66; *a,* evolute side view; *b,* edge view; *c,* oblique view of involute side; 80X. (=*Discorbis orientalis*).

PROTOZOA

       Sutures simple, lacking retral processes 2
2(1) Test planispiral, involute, symmetrical, aperture simple low arch at base
          of apertural face .... *Nonion*
       [*N. boueanum* (Pl. 8, Fig. 1)]
       Test somewhat asymmetrical with chambers extending farther toward
          umbilicus on one side than the other 3
3(2) Test evolute on one side; involute on other side with terminal chamber
          forming a lobe over the umbilicus .... *Nonionella*
       [see Pl. 14*b*]
       Test partly involute on one side; entirely so on the other side but with
          no umbilical lobe................................. *Pseudononion*
       [*Pseudononion japonicum* (Pl. 8, Fig. 2)]

Family **Camerinidae**

The Camerinidae possess a calcareous, perforate test with a planispiral, bilaterally symmetrical plan of growth. They tend to develop large lenticular tests.

Cushman recognizes the genera *Operculina* and *Operculinella*, found in Hawaii, and *Nummulites* (given as *Camerina*), an extinct fossil genus not found in Hawaii in Cushman's restricted sense. Cole (in Loeblich and Tappan, 1964) lumps all three into *Nummulites*. Cushman's usage is followed here.

KEY TO COMMON GENERA OF THE FAMILY CAMERINIDAE

1   Test planispiral, involute, at least in young........................... 2
       Test planispiral, evolute 3
2(1) Secondary skeleton with complex interior canal system, extinct, not
          found in Hawaii .... *Nummulites*
       Chambers simple....................................... *Operculinella*
       [*O. cumingii* (Pl. 7, Fig. 9)]
3(1) Chambers always simple .... *Operculina*
       [*O. philippinensis* (Pl. 7, Fig. 7)]
       Later chambers subdivided into chamberlets .... *Heterostegina*
       [*H. suborbicularis* (Pl. 7, Fig. 8)]

Family **Heterohelicidae**

The Heterohelicidae are a miscellaneous collection of genera of probably diverse origins. The distinguishing features are a calcareous perforate test with an initially planispiral plan of growth becoming biserial or uniserial. The supposed spiral beginning is much reduced in some genera and in some instances is questionable. *Bolivinella* is the only heterohelicid common in shallow Hawaiian waters. Cushman (1948) records a planispiral beginning in the microspheric forms, but this has been disputed by other authors, and the genus has been variously classified in several families. Later growth is biserial with long gracefully recurving chambers; overall test is diamond-shaped, much flattened (*B. folia* (Pl. 8, Fig. 6)).

Family **Buliminidae**

The Buliminidae are characterized by a calcareous perforate test, a serial plan of growth, and an aperture which is either loop-shaped, round on a short neck, or cribrate.

KEY TO COMMON GENERA OF THE FAMILY BULIMINIDAE

1. Plan of growth triserial, at least in young of microspheric forms 2
   Initial plan of growth other than triserial 7
2(1) Test sharply triangular in cross section (at least in young) 3
   Test rounded in cross section at least in young; aperture loop-shaped or cribrate or subtriangular; fusiform; aperture rounded on short neck 5
3(2) Test triserial throughout .... *Reussella*
   [*R. simplex* (Pl. 9, Fig. 1); *Reussella* cf. *R. aequa* (Pl. 9, Fig. 2); *R. spinulosa* (Pl. 9, Fig. 3)]
   Test becoming uniserial 4
4(3) Triangular cross section throughout; adult uniserial; aperture cribrate .... *Chrysalidinella*
   [*C. dimorpha* (Pl. 9, Fig. 4)]
   Adult flattened; chambers spreading, fan-shaped; apertures multiple openings along peripheral margin .... *Pavonina*
   [*P. flabelliformis* (Pl. 9, Figs. 5, 6)]
5(2) Triserial throughout; rounded section; fusiform .... *Uvigerina*
   [*U. porrecta* (Pl. 8, Fig. 12)]
   Becoming uniserial or loosely coiled 6
6(5) Subtriangular in section; triserial, later becoming loosely coiled or uniserial .... *Trifarina*
   [*Trifarina* sp. (Pl. 8, Fig. 13); *T. bradyi* (Pl. 8, Fig. 14)]
   Round in section; becoming uniserial in adult .... *Siphogenerina*
   [*S. raphana* (Pl. 8, Fig. 15)]
7(1) Plan of growth elongate spiral; many chambers per whorl ...... *Buliminella*
   [*B. milletti* (Pl. 8, Fig. 7)]
   Biserial, at least in young 8
8(7) Biserial throughout, aperture loop-shaped .... *Bolivina*
   [*B. striatula* (Pl. 8, Fig. 8); *B. compacta* (Pl. 8, Fig. 9)]
   Becoming uniserial in adult .... *Loxostomum*
   [*L. limbatum* (Pl. 8, Fig. 10)]

Fissurina is a genus of uncertain affinities. It is single-chambered, flattened, and has a slit aperture and an internal tube [*F. marginata* (Pl. 8, Fig. 11)]. It is similar in some respects to *Lagena* but has been classified with the Buliminidae because of the internal tube.

Family **Spirillinidae**

The Spirillinidae possess a calcareous perforate test and a planispiral evolute plan of growth consisting of a proloculus and a long undivided second chamber. Two genera are common. *Spirillina* possesses the characters of the family [see Pl. 14*d*]; *Planispirillina* is similar but secretes secondary nodes that obscure the whorls on one side [*P. denticulogranulata* (Pl. 9, Fig. 7)].

Family **Discorbidae**

The Discorbidae are calcareous and perforate with a trochoid growth plan, an open umbilical region, and a ventral aperture which does not extend beyond the periphery. The apertural area contains flaps projecting from the terminal chamber into the umbilicus forming more or less complicated apertures which serve to distinguish the genera. Common Hawaiian species tend to be planoconvex with the involute ventral side flattened.

KEY TO COMMON GENERA OF THE FAMILY DISCORBIDAE

1   Test height approaches or exceeds test diameter; over-all conical shape; chambers elongate, overlapping; aperture radial slit with a flap projecting over the open umbilicus .... *Neoconorbina* [*N. patelliformis* (Pl. 10, Figs. 1, 2)]
    Test height less than test diameter, usually less than half test diameter; sutural slits mark position of previous apertures 2
2   Umbilical region open; apertural flap constricts portion of the aperture forming two communicating parts .... *Rosalina* [*R.* cf. *R. vilardeboana* (Pl. 9, Fig. 8); *R. orientalis* (Pl. 9, Fig. 9)]
    Umbilical region covered by apertural flaps and filled to a greater or lesser extend by an apertural plug .... *Discorbis* [*D. mirus* (Pl. 10, Fig. 3)]

Family **Rotaliidae**

Various genera with a calcareous, perforate test and a trochoid growth plan are placed in the Rotaliidae. Most rotalids occur in deeper water; however, certain species of a few genera are found in shallow water.

PLATE 10
PROTOZOA
Family **Discorbidae** (in part)

Figure 1.—*Neoconorbina patelliformis* (Brady). BPBM no. A140; *a*, side view; *b*, end view; 100X. These are two individuals plastogamically fused. The flush cresent-shaped chambers are not distinct in this scanning electron micrograph and have been artificially outlined, in part, on the larger specimen.

Figure 2.—*Neoconorbina patelliformis* (Brady). BPBM no. A141; *a*, side view; *b*, involute end view; 100X. (=*Discorbina patelliformis*, =*Discorbis patelliformis*, =*Glabratella patelliformis*). The arrow points to the elongate radial slit aperture.

Figure 3.—*Discorbis mirus* (Cushman). BPBM no. A142; *a*, evolute side view; *b*, edge view; *c*, involute side view; 80X.

Family **Anomalinidae** (in part)

Figure 4.—*Cibicides* sp. BPBM no. A43; *a*, evolute side view; *b*, edge view; *c*, involute side view; 80X. The last four chambers were secreted under adverse conditions and are deformed.

Family **Rotaliidae**

Figure 5.—*Poroeponides cribrorepandus* Asano and Uchio. BPBM no. A63; *a*, evolute side view; *b*, edge view; *c*, involute side view; 40X.

Figure 6.—*Ammonia beccarii tepida* (Cushman). BPBM no. A65; *a*, evolute side view; *b*, edge view; *c*, involute side view; 100X. (=*Rotalia beccarii tepida*, =*Streblus beccarii tepida*).

Figure 7.—*Siphoninoides echinatus* (Brady). BPBM no. A16; *a*, apertural view; *b*, side view; 100X. (=*Planorbulina echinata*, =*Siphonina echinata*).

Family **Calcarinidae**

Figure 8.—*Calcarina* (?) *murrayi* (Heron-Allen and Earland). BPBM no. A143; *a*, evolute side view; *b*, edge view; *c*, involute side view; 100X. (=*Rotalia murrayi*). This species is of uncertain generic affinities, but seems to have secondary calcite secretions on the surface. It is a minute species, unlike other species of *Calcarina* which are large.

PROTOZOA 41

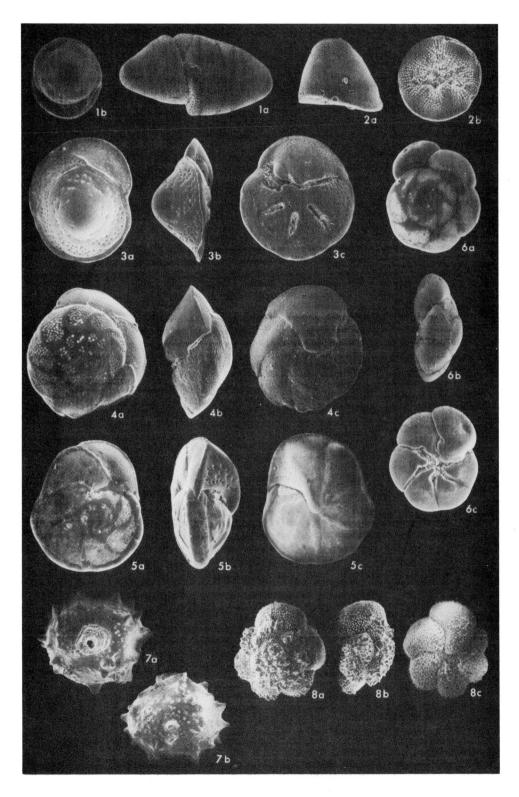

KEY TO COMMON GENERA OF THE FAMILY ROTALIIDAE

1 Plan of growth loosely trochoid; over-all test globular; aperture round and on short neck .... *Siphoninoides*
   [*S. echinatus* (Pl. 10, Fig. 7)]
   Plan of growth distinctly trochoid; over-all shape generally biconvex or planoconvex 2

2(1) Umbilical region filled with plug subdivided by open fissures; aperture an open slit along margin of last-formed chamber on involute side .... *Ammonia*
   [*A. beccarii tepida* (Pl. 10, Fig. 6)]
   Umbilical area indented but closed; primary aperture a slit parallel to axis of coiling extending from umbilicus to margin; secondary apertures in form of large pores scattered over apertural face .... *Poroeponides*
   [*P. cribrorepandus* (Pl. 10, Fig. 5)]

Family **Amphisteginidae**

The Amphisteginidae are similar to the Rotaliidae in possessing a calcareous, perforate test with a trochoid plan of growth but differ in possessing supplementary chambers on the involute ventral side. *Amphistegina* is the only Hawaiian genus, but it is one of the largest and most abundant forms on Hawaiian beaches. The juvenile trochoid plan of growth tends to become partly involute in the adult as chambers on the evolute dorsal side progressively overlap earlier chambers. The portion of the ventral side adjacent to the aperture is papillate [*A. madagascariensis* (Pl. 11, Figs. 1, 2)].

Family **Calcarinidae**

The Calcarinidae are similar to the Rotaliidae in possessing calcareous, perforate tests and a trochoid plan of growth, at least in the young. Simple forms are gradational with the Rotaliidae, but more complex forms differ by adding secondary calcite to the test and by forming large blunt spines and possessing internal pillars

---

PLATE 11
PROTOZOA
Family **Amphisteginidae**

Figure 1.—*Amphistegina madagascariensis* d'Orbigny. BPBM no. A144; *a*, evolute side view; *b*, edge view; *c*, involute side view; 40X. Sutures are flush and are not visible in this scanning electron micrograph, but are very evident under light microscopic examination.
Figure 2.—*Amphistegina madagascariensis* d'Orbigny. BPBM no. A145; *a*, evolute side view; *b*, edge view; *c*, involute side view; 40X.

Family **Cymbaloporidae**

Figure 3.—*Cymbaloporetta squammosa* (d'Orbigny). BPBM no. A146; *a*, evolute side view; *b*, edge view; *c*, involute side view; 80X (=*Rosalina squammosa*).
Figure 4.—*Cymbaloporetta bradyi* (Cushman). BPBM no. A67; *a*, evolute side view; *b*, edge view; *c*, involute side view; 40X. (=*Cymbalopora poeyi* var. *bradyi*).
Figure 5.—*Tretomphalus bulloides* (d'Orbigny). BPBM no. A147; *a*, evolute side view and *b*, edge view showing small coiled portion and large final balloon chamber, 80X; *c*, detail of balloon chamber showing coarse pores and yet coarser apertural pores, 300X. (=*Rosalina bulloides*).

PROTOZOA

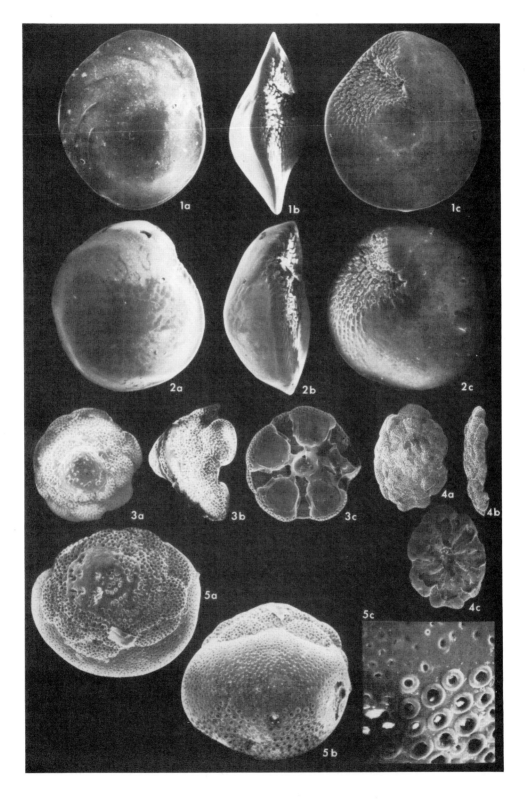

and canals. The Calcarinidae tend to be ultratropical. They are essentially absent from Hawaii but elsewhere are among the most common tropical genera.

Calcarina (?) *murrayi* (Pl. 10, Fig. 8) is a tiny species that occurs in Hawaii. It has often been referred to *"Rotalia"* in quotation marks for lack of a better designation, but is not a true *Rotalia*. Secondary calcite deposits revealed by the scanning electron microscope indicate possible affinities with *Calcarina. C.* (?) *murrayi,* however, is much smaller than other species placed in *Calcarina.*

Family **Cymbaloporidae**

Species assigned to the Cymbaloporidae are common and widespread near shallow tropical reefs, but the taxonomic relations of the family are poorly understood. Several named genera appear to be only growth stages in the life cycle of other genera. A number of named species within some genera appear to be morphological variants of one or a few species. Todd (1971) reviews the relations of the species but concludes that insufficient evidence is currently at hand to classify them into natural taxonomic units. Cushman's classification, although inadequate, is here utilized for lack of a better alternative. Plan of growth is similar to that of the Discorbidae in the young but becomes annular in the adult. Apertures are multiple and consist of numerous large pores variously arranged. The test is calcareous and perforate.

KEY TO COMMON GENERA OF THE FAMILY CYMBALOPORIDAE

1   Adult with large globular float chamber covering umbilical region; indistinguishable from other genera if float chamber missing .... *Tretomphalus*
    [*T. bulloides* (Pl. 11, Fig. 5)]
    No float chamber in any portion of life cycle 2

PLATE 12
PROTOZOA
Family **Cassidulinidae**

Figure 1.—*Cassidulina minuta* Cushman. BPBM no. A148; *a,* side view; *b,* edge view; 200X.
Figure 2.—*Cassidulina delicata* Cushman. BPBM no. A50; *a,* side view; *b,* edge view; 100X.

Family **Globigerinidae**

Figure 3.—*Candeina nitida* d'Orbigny. BPBM no. A149; *a,* evolute side view; *b,* edge view; 100X.
Figure 4.—*Globigerina eggeri* Rhumbler. BPBM no. A150; *a,* evolute side view; *b,* edge view; *c,* involute side view, 80X.
Figure 5.—*Orbulina universa* d'Orbigny. BPBM no. A21; 80X.
Figure 6.—*Globigerinoides sacculifer* (Brady). BPBM no. A151; *a,* evolute side view; *b,* edge view; *c,* involute side view; 80X. (=*Globigerina sacculifer).* This is a juvenile specimen that has not yet developed the elongate pointed final chamber.
Figure 7.—*Globigerinoides conglobatus* (Brady). BPBM no. A68; *a,* evolute side view; *b,* edge view; *c,* involute side view; 80X. (=*Globigerina conglobata).* The chambers are somewhat flattened and the sutures are deeply incised as fits *G. conglobatus.* The supplementary apertures are very obscure and may be lacking.

Family **Globorotaliidae**

Figure 8.—*Globorotalia* sp. BPBM no. A152; *a,* evolute side view; *b,* edge view; *c,* involute side view; 100X. A juvenile.
Figure 9.—*Globorotalia menardi* (d'Orbigny). BPBM no. A25; *a,* evolute side view; *b,* edge view; *c,* involute side view; 40X. (=*Rotalia menardii,* =*Pulvinulina menardii).*

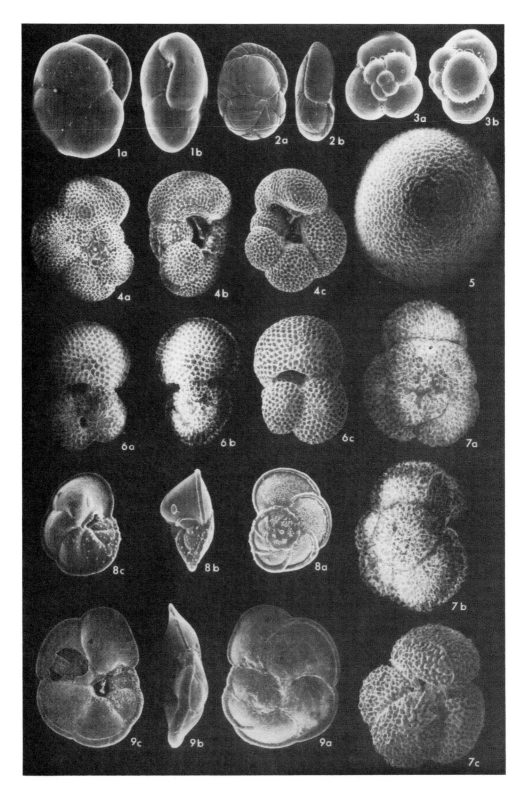

2(1) Test conical; umbilical area covered by plate .... *Cymbaloporetta*
[*C. bradyi* (Pl. 11, Fig. 4); *C. squammosa* (Pl. 11, Fig. 3)]
  Test compressed; final chambers in adult vertical to base ... *Cymbaloporella*
[see Pl. 14*c*]

Family **Cassidulinidae**

This family is characterized by a calcareous, perforate test, a peculiar biserially enrolled plan of growth, and a slit aperture elongated in the plane of coiling. The genus *Cassidulina* has the characters of the family without further special modifications. In general, it prefers cold water, but several of its species occur in shallow tropical reef areas [*C. minuta* (Pl. 12, Fig. 1); *C. delicata* (Pl. 12, Fig. 2)].

Family **Globigerinidae**

The most distinguishing features of the family are the globular chambers and over-all globular test shapes which reflect the planktonic mode of life of all genera within the family. The plan of growth, at least in juveniles, is a low trochospiral, usually with a large open umbilicus. Chambers may become embracing in adults, and in the morphological end-form known as *Orbulina,* a single spherical chamber covers the earlier trochoid growth stage. *Orbulina* has been considered a genus but is polyphyletic as evidenced by early growth stages of different kinds occurring within the test. The tests of the Globigerinidae are calcareous and perforate; the perforations are often quite large. Many genera possess long spines, which fall off upon death of the organism. Many species within the family are truly pelagic and are found in shallow near-shore waters only when blown in by storms. Certain species do inhabit near-shore waters, however.

PLATE 13
PROTOZOA
Family **Anomalinidae**

Figure 1.—*Anomalina glabrata* Cushman. BPBM no. A153; *a,* evolute side view; *b,* edge view; *c,* involute side view; 100X.

Figure 2.—*Anomalina* sp. BPBM no. A154; *a,* evolute side view; *b,* edge view; *c,* involute side view; 80X.

Figure 3.—*Cibicides lobatulus* (Walker and Jacob). BPBM no. A155; *a,* evolute side view; *b,* edge view; *c,* involute side view; 80X. *(=Nautilus lobatulus, =Truncatulina lobatula).* Authors have figured forms similar to this as *C. lobatulus.* It is an attaching species that assumes many morphological variations.

Figure 4.—*Cibicides lobatulus* (Walker and Jacob). BPBM no. A156; *a,* evolute side view; *b,* edge view; *c,* involute side view; 80X.

Family **Planorbulinidae**

Figure 5.—*Gypsina globula* (Reuss). BPBM no. A19; *a,* top view; *b,* edge view; 40X. This has traditionally been considered a valid genus, but Nyholm, as mentioned by Todd (1965), considers it a resting stage in the life cycle of *Cibicides.*

Family **Homotremidae**

Figure 6.—*Miniacina miniacea* (Pallas). BPBM no. A157; *a,* side view, and *b,* top view, 20X; *c,* detail of one branch showing terminal apertures, 100X; *d,* detail of surface showing fine pores and larger surficial aperture (pillar pore), 1000X. *(=Millepora miniacea, =Polytrema miniaceum).* The arrows point to siliceous sponge spicules that have been picked up by the animal and incorporated into the test wall at the ends of the branches. This habit is typical of the Family Homotremidae.

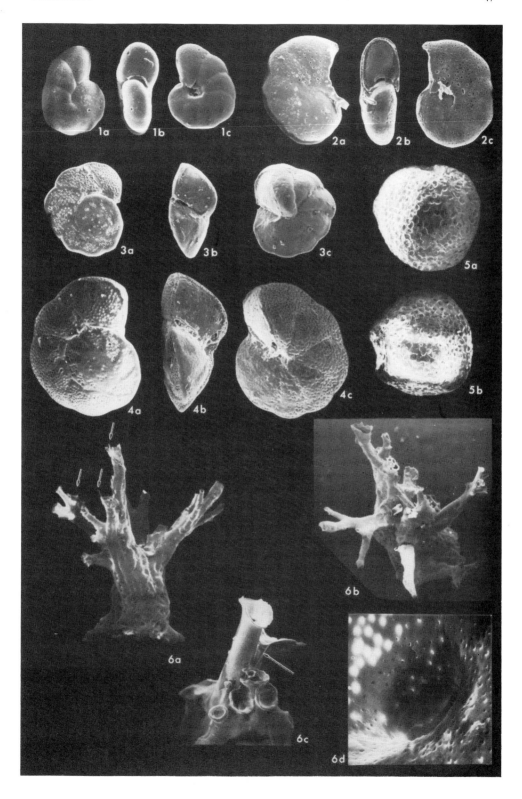

KEY TO COMMON GENERA OF THE FAMILY GLOBIGERINIDAE

1   Single spherical chamber visible from exterior of adult; apertures numerous large pores scattered over surface .... *Orbulina*
    [*O. universa* (Pl. 12, Fig. 5)]
    Several chambers visible from exterior of adult 2
2(1) Aperture single large umbilical opening; many long slender spines project from test of living animal .... *Globigerina*
    [*G. eggeri* (Pl. 12, Fig. 4)]
    Apertural openings multiple along sutures 3
3(2) Several large supplementary apertures along sutures; large primary umbilical aperture; many fine spines project from test of living animal .... *Globigerinoides*
    [*G. sacculifer* (Pl. 12, Fig. 6); *G. conglobatus* (Pl. 12, Fig. 7)]
    Adult with numerous small apertural openings along sutures; surface smooth; no spines .... *Candeina*
    [*C. nitida* (Pl. 12, Fig. 3)]

Family **Globorotaliidae**

The Globorotaliidae have a calcareous, perforate test and a trochoid plan of growth. They are planktonic and were derived from the Globigerinidae by the development of a flattened dorsal side giving them a rotalid shape. Chambers are inflated. The genera are largely pelagic, but empty tests of *Globorotalia* are occasionally found in near-shore sands. This genus has the characters of the family and usually possesses a peripheral keel [*Globorotalia* sp. (Pl. 12, Fig. 8); *G. menardi* (Pl. 12, Fig. 9)]. Some authors consider species without the keel to represent another genus.

Family **Anomalinidae**

The genera of this family possess a calcareous, perforate test, a trochoid plan of growth, and all shallow Hawaiian genera have a slit aperture which extends from the ventral side across the periphery. Two genera are commonly found. *Anomalina* has the characters of the family. It often tends to become partly involute on the dorsal side as well as on the ventral side and tends toward a biconvex planospiral test [*A. glabrata* (Pl. 13, Fig. 1); *Anomalina* sp. (Pl. 13, Fig. 2)]. *Cibicides* is planoconvex and is often attached by the flat dorsal side. Some authors consider species that do not attach to be different genera. The aperture usually extends across the periphery and along the dorsal side for most of the length of the final chamber [*Cibicides lobatulus* (Pl. 13, Figs. 3, 4); *Cibicides* sp. (Pl. 10, Fig. 4)].

Family **Planorbulinidae**

The Planorbulinidae have a calcareous, perforate test. In juveniles the plan of growth is similar to that of the Anomalinidae. Adults develop various modified plans of growth. *Planorbulina* [see Pl. 14e] becomes annular and develops two apertures per chamber, each of which is covered by a new chamber with two apertures. The flattened test with annular chambers is superficially similar to a few other genera, particularly to *Cymbaloporetta*. The juvenile plan of growth and nature of apertures will distinguish one from another; however, *Gypsina* produces adult chambers that grow on top of one another so that an irregularly spherical mass is formed. The coarse pores serve as apertures [*G. globula* (Pl. 13, Fig. 5)].

Family **Homotremidae**

The three peculiar genera in this family bear little superficial resemblance to other foraminifers; they are often mistaken for bryozoans. Although the juveniles have a typical trochoid plan of growth, the chambers soon become permanently attached to the substrate and grow upward in an irregularly branching mass which obscures the early growth. The genera are also unusual in possessing distinctive colorations of taxonomic significance. They are very common attached organisms, but empty tests are rarely found in sand accumulations. This group is well treated by Hickson (1911).

KEY TO THE GENERA OF THE FAMILY HOMOTREMIDAE

1   Test with slender branching projections; surface finely perforate; larger open apertures; light red color .... *Miniacina* [*M. miniacea* (Pl. 13, Fig. 6)]
    Surface not perforate in adult other than by apertures; branches short, stubby 2

2(1) Large apertures covered by perforated plates; dark red color.... *Homotrema*
    Apertures not covered by plates; orange red color .... *Sporadotrema*

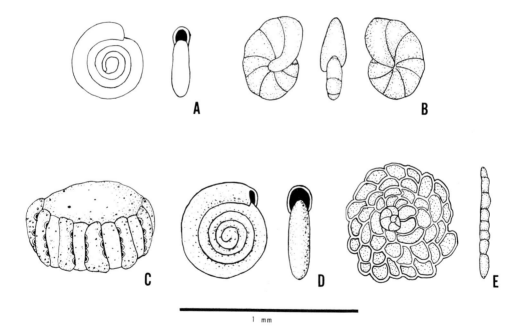

PLATE 14
PROTOZOA

The genera are included in the key to Foraminifera, but are not illustrated on Plates 2 through 13. The drawings are stylized and do not represent particular specimens. *a, Cornuspira; b, Nonionella; c, Cymbaloporella; d, Spirillina; e, Planorbulina.*

GLOSSARY (PROTOZOA)

aperture: large opening or multiple openings from chamber to exterior.
biloculine growth plan: 2 chambers externally visible; chambers added at 180° to one another and are involute.
biserial growth plan: elongate growth plan with chambers in two parallel rows.
collectoderm: material secreted by a folliculinid to cement its lorica to the substrate.
cribrate aperture: aperture sievelike, of numerous small holes.
dorsal: for trochoid plan of growth, the evolute side.
evolute: younger chambers do not cover older chambers; all chambers back to proloculus are visible.
foramina: opening; opening between chambers; may be previous aperture or may be formed by secondary solution of test wall.
fusiform: rice-grain-shaped; circular in cross section; elongate; largest diameter near middle and tapering toward each end.
hypotype: or plesiotype, a specimen referred to a previously described species, and used as the basis of a new description or figure of that species.
involute: young chambers cover older chambers; only the last whorl of chambers is visible from the exterior.
lorica: protective shell formed by several ciliate protozoan groups, including folliculinids.
moniliform nucleus: nucleus composed of a series of granular masses loosely strung together; resembles a string of beads.
pelagic: strictly pertaining to open sea as opposed to coastal waters; sometimes loosely and incorrectly used as equivalent to "planktonic."
peristomal lobes: a characteristic feature of folliculinids. The sides of the buccal cavity (peristome) are separated by constrictions of the protoplasm into two segments termed peristomal lobes.
planktonic: floating, carried by currents.
planispiral growth plan: chamber or chambers coiled in a single plane; may be evolute or involute.
plastogamy: type of sexual reproduction in which 2 adult tests fuse together along their dorsal surfaces.
pore: minute opening; distributed over the surface of the test of certain groups of forams.
proloculus: initial chamber in foram test.
protoplast: the living portion of a protozoan, as distinguished from any nonliving shell material which may also be present.
pseudochitin: a flexible proteinaceous material similar to keratin making up the test wall or part of the wall of some groups.
pseudopod: temporary protoplasmic projection used for feeding, locomotion, and attachment.
quinqueloculine growth plan: typical of at least early portion of most genera in the Family Miliolidae; each chamber is ½ of the test circumference; aperture is at opposite end of test to that of preceding chamber; chambers added about a common elongate axis at angles of 144° so that a cycle is completed in two revolutions; usually 5 chambers externally visible, at angles of 72° to one another.
radiate aperture: aperture consisting of elongate slits radially arranged.
spherical nucleus: a compact nucleus forming a spherical to subspherical mass.
spiral whorl: structural feature which may be present on the neck of a folliculinid lorica; usually in the form of a helical groove.
test: a shell covered by a layer of living material; present in Foraminifera.
triloculine growth plan: 3 chambers, externally visible, coiled about an elongate axis; each chamber is ½ of the test circumference; aperture is at opposite end of test to that of preceding chamber; chambers added at angles of 120° to one another so that a complete cycle is completed in one revolution.
triserial growth plan: elongate spiral growth plan in which chambers grow at an angle of 120° to one another, 3 chambers forming a complete whorl.
trochoid growth plan: test is evolute on one side showing all chambers, but is involute on the other side and shows only the last-formed whorl.
ventral: for trochoid plan of growth, the involute side.

REFERENCES (PROTOZOA)

Andrews, E. A.
    1944. A Folliculinid from the Hawaiian Islands. *Trans. American Microscopical Soc.* 63:321-325.

Arnold, Z. M.
    1954. Culture Methods in the Study of Living Foraminifera. *J. Paleontology* 28: 404-416.
    1974. Techniques for the Study of Living Foraminifera. In R. H. Hedley and C. G. Adams (eds.), *Foraminifera,* Vol. 1, pp. 153-206. London: Academic Press.

Bagg, R. M., Jr.
    1908. Foraminifera Collected near the Hawaiian Islands by the Steamer *Albatross* in 1902. *Proc. U.S. National Mus.* 34: 113-172. 1 pl.

Barker, R. W.
    1960. *Taxonomic Notes on the Species Figured by H. B. Brady in His Report on the Foraminifera Dredged by H. M. S.* Challenger *during the Years 1873-1876.* Soc. Economic Paleontologists and Mineralogists Spec. Pub. 9. 238 pp., 115 pl. Tulsa, Oklahoma.

Brady, H. B.
    1884. Report on the Foraminifera Dredged by H. M. S. *Challenger* during the Years 1873-1876. *Report of the Scientific Results of the Voyage of H. M. S.* Challenger, *Zoology,* Vol. 9. 814 pp., 115 pl.

Cushman, J. A.
    1910-1917. *A Monograph of the Foraminifera of the North Pacific Ocean.* Smithsonian Inst. Bull. 71, pts. 1-6. 664 pp., 136 pls., 473 figs.
    1924. *Samoan Foraminifera.* Carnegie Inst. Washington Publ. 342. 75 pp., 25 pls.
    1925. Foraminifera of the Tropical Central Pacific. In *Marine Zoology of Tropical Central Pacific,* pp. 121-144. B. P. Bishop Mus. Bull. 27. Honolulu.
    1932-1942. *The Foraminifera of the Tropical Pacific: Collections of the* Albatross, *1899-1900.* U.S. National Mus. Bull. 161, pts. 1-3. (Part 4 by Todd, 1965.)
    1948. *Foraminifera: Their Classification and Economic Use.* Cambridge: Harvard Univ. Press. 605 pp.

Edmondson, C. H.
    1946. *Reef and Shore Fauna of Hawaii.* B. P. Bishop Mus. Spec. Publ. 22. Honolulu. 381 pp.

Hickson, Sydney J.
    1911. On *Polytrema* and Some Allied Genera. *Trans. Linnean Soc. London,* 2nd Ser., Zoology 14:443-462. Pls. 30-32.

Honigberg, B. M., and others
    1964. A Revised Classification of the Phylum Protozoa. *J. Protozoology* 11:7-20.

Loeblich, A. L., and H. Tappan
    1964. Sarcodina, Chiefly "Thecamoebians" and Foraminiferida. *Treatise on Invertebrate Paleontology,* pt. C. 900 pp.

Matthews, D. C.
    1953. New Hawaiian Records of Folliculinids (Protozoa). *Trans. American Microscopical Soc.* 72: 344.
    1962. Additional Records of Folliculinids (Protozoa) in Hawaii. *Pacific Science* 16(4): 429-433.
    1963. Hawaiian Records of Folliculinids (Protozoa) from Submerged Wood. *Pacific Science* 17(4): 438-443.
    1964. Recent Observations on Neck Extensions in Folliculinids (Protozoa). *Pacific Science* 18(2): 229-235.
    1968. The Folliculinids (Protozoa) of Ago Bay, Japan, and Their Relation to the Epifauna of the Pearl Oyster *(Pinctada martensii). Pacific Science* 22(2): 232-250.

Moberly, R., Jr., L. D. Baver, Jr., and A. Morrison
    1965. Source and Variation of Hawaiian Littoral Sand. *J. Sedimentary Petrology* 35: 589-598.

Muller, P. M.
    1974. Sediment Production and Population Biology of the Benthic Foraminifer *Amphistegina madagascariensis. Limnology and Oceanography* 19(5): 802-809.

Murray, J. W.
    1973. *Distribution and Ecology of Living Benthic Foraminiferids.* New York: Crane, Russak. 274 pp.

Phillips, F. J.
    1973. Sandy Subtidal Community (in part). Marine Environment Impact Assessment Report for Hawaiian Electric Company, Incorporated, Kahe Point Facility, Oahu, Hawaii, pp. 5-19 to 5-38. URS Research Co. Document URS 7220-3. Figs. 5-3 to 5-31. (Unpublished.)

Resig, J. M.
    1969. *Paleontological Investigations of Deep Borings on the Ewa Plain, Oahu, Hawaii.* Hawaii Inst. Geophysics, Pub. HIG-69-2. 99 pp., 9 pls.
    1974. Recent Foraminifera from a Landlocked Hawaiian Lake. *J. Foraminiferal Res.* 4(2): 69-76.

Rhumbler, L.
    1906. Foraminiferen von Laysan und den Chatham-Inselen. *Zoologische Jahrbuecher Abteilung Systematik* 24: 21-80.

Ross, C. A.
    1972. Biology and Ecology of *Marginopora vertebralis* (Foraminiferida), Great Barrier Reef. *J. Protozoology* 19(1): 181-192.

Röttger, R.
    1972. Die Kultur von *Heterostegina depressa* (Foraminifera: Nummulitidae). *Marine Biology* 15: 150-159.

Todd, Ruth
    1961. Foraminifera from Onotoa Atoll, Gilbert Islands. *U.S. Geological Survey Professional Pap.* 354-H, pp. i-iii, 171-192. Pl. 22-25.
    1965. *The Foraminifera of the Tropical Pacific Collections of the Albatross, 1899-1900.* U.S. National Mus. Bull. 161, pt. 4. (Parts 1-3 by Cushman, 1932-1942.)
    1971. *Tretomphalus* (Foraminifera) from Midway. *J. Foraminiferal Res.* 1(4): 162-169.

Walker, D. A., A. E. Linton, and C. T. Schafer
    1974. Sudan Black B: A Superior Stain to Rose Bengal for Distinguishing Living from Nonliving Foraminifera. *J. Foraminiferal Res.* 4(4): 205-215.

PORIFERA

PATRICIA R. BERGQUIST
University of Auckland

THE PORIFERA (sponges) are a clearly defined group of predominantly marine animals. They are common around most shorelines and are important constituents of most bottom faunas.

The position of sponges in zoological classification is at the base of the multicellular organisms (Metazoa). Porifera differ from Protozoa in having a distinctly cellular construction, and from other metazoans in lacking well-formed tissues. There are many features of organization and physiology that are peculiar to sponges, but the most important is their possession of an elaborate filter-feeding apparatus based on a complex system of pores and canals and powered by the asynchronous action of peculiar flagellate cells termed choanocytes. Support for the soft structures of the sponge body is provided by discrete calcareous or siliceous spicules or by spongin fibers, either separately or in combination with spicules.

Adult sponges are usually sessile; the motile phase of the life history is the larva, produced as a result of sexual reproduction. This larva is short-lived and soon settles and begins development. It may be a hollow flagellated amphiblastula or a solid ciliated stereogastrula. Asexual reproduction is common among sponges and is achieved either by the production of buds, as in *Tethya,* or of gemmules, as in most freshwater forms and such marine forms as *Haliclona.*

The primary systematic division of the phylum into three classes is based on the composition and morphology of the skeleton:

Class CALCAREA: a skeleton of discrete spicules composed of calcium carbonate.
Class DEMOSPONGIAE: a skeleton of siliceous spicules or of spongin fibers. In some few cases, no skeleton is present. The pattern of these spicules is either monactinal, diactinal, triactinal (occasionally), or tetractinal. The Demospongiae are the most numerous, diverse, and important group in most habitats.
Class HEXACTINELLIDA: a skeleton of hexactinal (triaxon) siliceous spicules. These forms are usually confined to deep water.

SPONGE STUDY TECHNIQUES

Because of the notorious variability of sponges in such features as over-all shape, coloration, and nature of the surface, it is always desirable to preserve the specimen in such a way that further study is possible. Most attributes of the sponge that are important in classification can be determined with relative ease from carefully preserved material.

Full color notes must be taken in the field, preferably making reference to some objective color standard (for example, Munsell, 1942). After study of the surface characteristics under low-power binocular magnification, gross fixation of the whole specimen should be in 95 percent ethyl alcohol, changed after 24 hours. It is advisable to preserve small portions of the sponge in standard histological fixatives.

In cases where only spongin (a fibrous protein) comprises the skeleton, as in the Dictyoceratida and Dendroceratida, sectioning is necessary to show the spongin characteristics. In the majority of Demospongiae one or many types of siliceous spicules are present, and these are best studied by maceration of the tissue in concentrated nitric acid. Boil a fragment of the siliceous sponge, taken so as to include both surface and interior components, in concentrated nitric acid in a fume-hood. (For calcareous sponges, use concentrated potassium hydroxide.) When the liquid clears or becomes only pale yellow, remove from the flame and decant into a centrifuge tube. Centrifuge in two changes of water, then in two changes of 95 percent alcohol. This preparation of clean spicules can then be well shaken and a small portion placed on a slide on a warming tray. When dry, mount in balsam, cover, and examine.

An alternative method is to macerate in sodium hypochlorite or in a commercial bleaching agent, such as Clorox. After disintegration in the sodium hypochlorite, the sample is centrifuged in three changes of water and two changes of 95 percent alcohol. The preparation is then decanted on to a slide, allowed to dry, and mounted in balsam.

Temporary preparations can be obtained by carrying out the maceration on a slide. After the tissue disintegrates, the preparation can be covered and examined. Permanent mounts should be prepared in a centrifuge tube.

Other skeletal elements can be observed in freehand sections taken with a razor blade. Staining is not essential, but if the sections are stained in a saturated solution of basic fuchsin in 95 percent alcohol, histological details and spongin elements, which stain bright, clear red, are easier to observe.

Such sections are best taken at right angles to the sponge surface. In some cases embedding the sponge fragment in paraffin facilitates sectioning. Material should be dehydrated beforehand and left six to twelve hours in paraffin. The paraffin must be removed before staining. These sections will show the location of various spicule types in the sponge and the degree of development of the spongin skeleton in relation to the mineral skeleton.

HAWAIIAN SPONGE FAUNA

Based on the taxonomic treatments by de Laubenfels (1950, 1951, 1954a, b, 1957) and Bergquist (1967), as well as on new information presented herein, some 63 species of Porifera are known from Hawaiian waters to a depth of 100 m. De Laubenfels in his works discussed the earlier records on Hawaiian sponges by Haeckel (1872), Lendenfeld (1910), and Edmondson (1946). The latter paper by Edmondson dealt with the asexual reproduction of *Tethya* cf. *diploderma* (as *Donatia deformis*).

Twenty-four species of shallow-water Porifera listed below are known only from the Hawaiian Islands. Of the remaining species there are several found as members of the fouling community exclusively. These include *Halichondria melanadocia, Tedania ignis, Mycale cecilia,* and *Zygomycale parishi* as reported by Bergquist (1967), all of which can be considered accidental introductions to Hawaii. Aside from the species reported only from Hawaii, the remainder are distributed for the most part in the western Pacific, Indo-Malaysian, and Australian regions, but also include representatives from the eastern Pacific or those with a pantropical range.

The keys that follow relate only to those shallow water species known from Hawaii at present.[1] It is expected, however, that many more forms remain to be described or recorded from the Hawaiian region, and the student of sponges must anticipate new material that will require a specialist's identification. The keys are based as far as possible on characteristics of living or freshly collected specimens before fixation. In addition, the type and size of skeletal elements are given, in brackets, for each species keyed, and many of these elements are defined and illustrated.

KEY TO HAWAIIAN SHALLOW-WATER DEMOSPONGIAE

1 Texture of sponge firm, compact, or hard 2
   Texture soft, easily compressible, or sponge so thin that no
      texture is discernable 10

2(1) Sponge of encrusting or spreading form 3
   Sponge usually subspherical or lamellate but may be
      merely massive 5

3(2) Microscleres present 4
   Microscleres absent; sponge pale brown in color with prominent,
      tangentially arranged spicules in dermis; rare .... *Petrosia puna*
   [Oxeas, styles, and strongyles 132–165 × 13–14 μm; oxeas
      96 × 14 μm][2]

4(3) Microscleres spirasters; sponge brick red to orange with dark
      layer below surface; common, particularly under ledges at
      mean low water .... *Spirastrella coccinea*
   [Tylostyles 360–500 × 6–12 μm; spirasters 10–50 μm]
   Microscleres oxyasters; sponge color yellow gray; texture
      cartilaginous; rare .... *Zaplethea digonoxea*
   [Oxeas 400–520 × 7–12 μm; microxeas 105 × 3 μm;
      oxyasters 10–20 μm]

5(2) Surface showing prominent pattern of grooves and tubercles;
      body usually spherical with attachment processes 6
   Surface not so patterned 7

6(5) Color externally black, internally yellow; cortical region very
      thin and not noticeably more dense than endosome;
      occasional .... *Tethya* sp.
   [Strongyles or strongyloxeas 340–980 × 7–20 μm;
      spherasters 20–64 μm; tylospherasters 4–10 μm]
   Color externally white to light brown, internally yellow; outer
      cortical region much more dense than deeper layers;
      common .... *Tethya* cf. *diploderma*
   [Strongyles or strongyloxeas 500–1500 × 8–20 μm;
      spherasters 25–75 μm; tylasters with microspined rays
      5–11 μm; oxyasters 16–25 μm]

---

[1] Several species in the list of Demospongiae have not been included in the key.
[2] 1 μm = 0.001 mm, also known as 1 micron (μ).

7(5) Surface very hispid; color externally gray, internally white;
  occasional...*Asteropus kaena*
  [Oxeas 640–110 × 3–36 μm; sanidasters 8–16 μm;
  oxyasters 20–40 μm]
  Surface smooth or granular in appearance 8
8(7) Form of sponge body a wavy lamella; texture hard; color
  brown .... *Leiodermatium* sp.
  [Desmas; skeleton extremely compacted]
  Form irregular, color white 9
9(8) Microscleres sterrasters, forming a cortical palisade;
  occasional.. *Geodia gibberella*
  [Oxeas 620 × 15 μm; plagiotriaenes 410 × 18 μm;
  sterrasters 37 μm; oxyasters 5 μm]
  No sterrasters among microscleres; surface layer compounded
  of megascleres; rare............................. *Stelletta debilis*
  [Oxeas 720 × 18 μm; plagiotriaenes 320 × 16 μm;
  oxyasters 10 μm]
10(1) Sponge surface elevated into prominent conules 11
  Sponge surface not conulose, but may be tuberculate or
  corrugated 19
11(10) Sponge extremely thin, characteristically occurring in small
  patches 12
  Sponge thick, spreading, or massive, often with long
  surface projections 15
12(11) Surface having a fine reticulation of sand grains; color deep
  purple to black .... *Aplysilla violacea*
  [Spongin fibers rising from a basal plate, not branching
  and containing no detritus]
  Surface lacking such reticulation 13

PLATE 1
PORIFERA

Figure 1.—Acanthostyles
Figure 2.—Anisochela
Figure 3.—Birotulate
Figure 4.—Centrotylote oxea
Figure 5.—Isochela
Figure 6.—Strongylote euaster
Figure 7.—Monolophotriaene
Figure 8.—Reticulation of spongin fibers
  cored with sand. A fiber is shown
  (right) extending into a surface conule.
Figure 9.— Oxeote euaster
Figure 10.—Desmas
Figure 11.—Oxeas
Figure 12.—Oxyaster

Figure 13.—Tornote
Figure 14.—Style
Figure 15.—Onychaetes
Figure 16.—Toxas
Figure 17.—Triact
Figure 18.—Sagittal triact
Figure 19.—Tylaster
Figure 20.—Tylospheraster
Figure 21.—Tylote
Figure 22.—Anatriaene
Figure 23.—Plagiotriaene
Figure 24.—Raphide
Figure 25.—Strongyle

PORIFERA

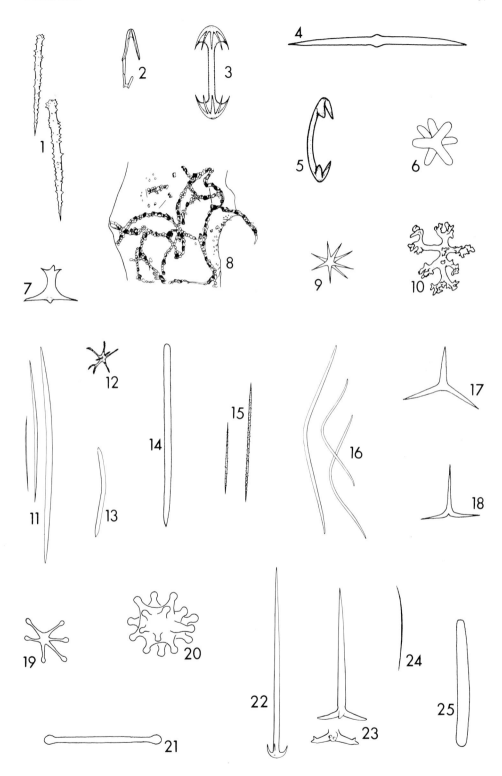

13(12) Sponge color extremely dense, blue or yellow 14
Sponge color translucent, rose pink .... *Dendrilla cactus*
[Commonly naked portions of dark fibers protrude from
surface; massive, irregular, or lobose shape]

14(13) Sponge blue .... *Pleraplysilla hyalina*
[Spongin fibers rising from a basal plate, not branching
and containing no detritus]
Sponge yellow .... *Aplysilla sulphurea*
[Upon dying, turns blue within minutes, then blackish purple]

15(11) Sponge surface membrane supporting a reticulation of spicule
fibers and stretched over the numerous conules and ridges;
external color variable—red brown, purple brown, dull
lavender, orange, or pale yellow; internal color dull
yellow; common .... *Zygomycale parishi*
[Styles 210–350 × 3–10 μ m; anisochelae 2 sizes—40–48 μ m
and 18–20 μ m; palmate isochelae 10 μ m; sigmas 2 sizes—
75–90 × 5 μ m and 25–30 × 1–2 μ m; toxas 30 μ m; raphides
10–15 × 1 μ m]
No reticulation visible on sponge surface; no mineral skeleton 16

16(15) Color bright sulphur yellow rapidly changing to greenish black
when removed from water and dark brick red in alcohol;
common .... *Psammaplysilla purpurea*
[Spongin fibers with a reticulate structure]
Color deep to pale purple, black, or gray 17

17(16) Sponge body a series of intersecting lamellae rising from an
encrusting base; color pale slate gray externally, dark
green internally; locally abundant................. *Dysidea herbacea*
[Spongin fibers filled with detritus]
Sponge body massive lumplike, over-all dimensions often as
small as 2 × 2 cm, but may be much larger in deep water 18

18(17) Fibrous skeleton forming a compact reticulum, very few fibers
contain detritus; color jet black; surface covered with
extremely fine conules; locally abundant .... *Spongia oceania*
[Spongin fibers of even diameter]
Fibrous skeleton not regularly arranged and all fibers packed
with detritus; color dull purple; surface studded with multiple
conules about 2 mm high; rare .... *Dysidea* sp.

19(10) Sponge boring in coral substrate; bright orange, usually visible at
inhalant and exhalant openings only; very
common .... *Cliona vastifica*
[Tylostyles 240–300 × 4–7 μ m; spined oxeas 75–85 × 4–5 μ m;
spirasters 8–18 × 2–3 μ m]
Sponge not boring 20

20(19) Sponge habit basically encrusting or spreading 21
Sponge massive, tubular or ramose 42

21(20) Sponge a very dense color—blue, red, yellow, or violet 22
       Sponge color either not one of the above or a pale translucent shade 30
22(21) Sponge with spiny spicules among megascleres 23
       Sponge having no spiny spicules 26
23(22) Sponge color red 24
       Sponge color blue 25
24(23) Spiny spicules acanthoxeas and acanthostyles, color vivid
           carmine red; sponge very thin; rare .... *Naniupi ula*
       [Styles 190 × 4 μm; acanthostyles 130 × 7 μm; acanthoxeas
           110 × 4 μm; isochelae 21 μm]
       Spiny spicules only acanthostyles; color bright clear red;
           common .... *Microciona* sp.
       [Large styles 200–350 × 9–14 μm; dermal styles
           220–294 × 2–4 μm; acanthostyles 80–110 × 5–7 μm;
           isochelae 11–13 μm; toxas 38–100 μm]
25(23) Sponge surface very hispid with spicules projecting up to 2 mm;
           color blue to greenish blue .... *Eurypon nigra*
       [Tylostyles 192–2000 × 6–10 μm; fine tylostyles 200–500 × 2.0
           μm; acanthostyles 90–160 × 6–9 μm]
       Sponge surface very smooth and skinlike, blue pigment with
           tendency to bleed and stain on contact .... *Hymedesmia* sp.
26(22) Megascleres tylostyles 27
       Megascleres not tylostyles 28
27(26) Sponge vivid bright yellow; subdermal channels very obvious;
           sponge exceedingly thin; rare .... *Prosuberites oleteira*
       [Tylostyles 230 × 7 μm]
       Sponge deep blue; a relatively thin crust 1 mm to 2 mm;
           exceedingly common .... *Terpios granulosa*
       [Tylostyles 180–360 × 2–6 μm (often with 4 lobed heads)]
28(27) Color red 29
       Color violet; megascleres oxeas, microscleres toxas; texture
           exceedingly soft; oscules sometimes elevated on small
           chimneys; common .... *Toxadocia violacea*
       [Oxeas 120–140 × 4–7 μm; toxas 60 × 1 μm]
29(28) Sponge surface slimy; oscules sometimes elevated above sponge surface;
           megascleres styles and tylotes; very common .... *Tedania ignis*
       [Spined tylotes 180–210 × 3–4 μm; styles 160–210 × 6–8 μm;
           onychaetes 200 × 1–2 μm]
       Sponge surface not slimy but crisp, slightly hispid; sponge 2 mm to 3 mm
           thick with prominent subdermal channels;
           common .... *Axociella kilauea*
       [Large styles 280–630 × 12–15 μm; tylostyles 144 × 2 μm; toxas
           30–60 μm; palmata isochelae 14 μm]
30(21) Sponge with prominent exhalant channels just below dermal membrane .. 31
       Sponge lacking prominent exhalant channels 34
31(30) Megascleres of a single type 32
       Megascleres of more than one type 33

32(31) Microscleres present, megascleres tylostyles; sponge a thin orange to gray crust with prominent surface channels; common under coral boulders .... *Spirastrella coccinea*
[Tylostyles 360–500 × 6–12 μm; spirasters 10–50 μm]
Microscleres absent, megascleres oxeas of several sizes; sponge a thick encrustation 1.0 cm to 1.5 cm; yellow to green yellow in color; locally abundant (Ala Moana) .... *Halichondria dura*
[Oxeas 650–680 × 20–27 μm]

33(31) Megascleres acanthostyles and styles; sponge a soft crust 1.5mm to 2.0 mm thick with punctate surface; variable colors ranging from gray through brown to orange; occasional .... *Microciona maunaloa*
[Large subtylostyles 425 × 13 μm; styles 240 × 9 μm; fine styles 120–160 × 6 μm; acanthostyles 42 × 8 μm; toxas 56–132 μm; palmate isochelae 14 μm and 5–6 μm]
Megascleres styles and tylotes; color dull orange; occasional.............................. *Tedania macrodactyla*
[Tylotes 165–210 × 2.5–4.0 μm; styles 150–200 × 3–7 μm; onychaetes 160–140 × up to 2.0 μm]

34(30) Sponge surface slimy 35
Sponge surface not slimy 37

35(34) Some megascleres triactinal; sponge olive brown to gray; rare .... *Plakortis simplex*
[Triact rays 100 × 5–7 μm; centrotylote oxeas]
All megascleres diactinal or monactinal............................ 36

36(35) Sponge white; megascleres oxeas and occasional strongyles, microscleres absent; rare .... *Rhaphisia myxa*
[Oxeas or strongyles of variable shape 85–150 × 1–6 μm]
Sponge orange; megascleres tylotes and oxeas, microscleres isochelae and sigmas; often occurring, when encrusting, upon colonies of green alga *Dictyosphaeria*; common .... *Damiriana hawaiiana*
[Tylotes 200 × 5 μm; oxeas 200 × 8 μm; arcuate isochelae 15–30 μm; sigmas 13 μm]

37(34) Megascleres styles, some microscleres anisochelae 38
No styles or anisochelae present 39

38(37) Dermal membrane containing areas of more dense or contrasting pigment, sponge soft but elastic; color variable—pink, lavender, yellow, orange, red, purple to dull blue; very common .... *Mycale cecilia*
[Styles 240–250 × 4–6 μm; 30–42 μm; anisochelae 15–24 μm]

PLATE 2
PORIFERA

Figure 1.—Quadriradiate
Figure 2.—Sigma
Figure 3.—Strongyloxea (*a*) entire; (*b*) variations of rounded end
Figure 4.—Spongin

Figure 5.—Spheraster
Figure 6.—Spirasters
Figure 7.—Sterraster
Figure 8.—Tylostyles

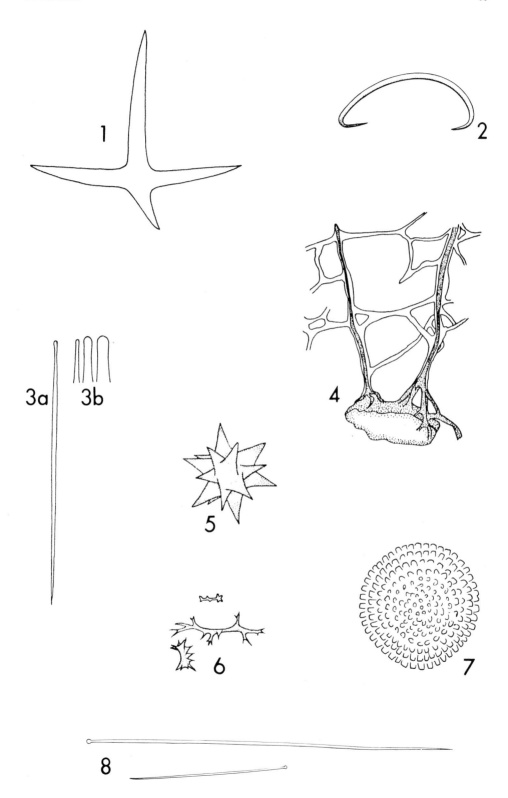

	Dermal membrane lacking contrasting pigment but supporting a reticulum of megascleres; color orange to pale gray; rare .... *Mycale (Carmia) contarenii* [Styles 350–400 × 10–13 μm; sigmas 80 × 5 μm; toxas 60 μm; anisochelae 2 sizes—40 μm and 10 μm]
39(37)	Megascleres strongyles 41
	Megascleres other than strongyles 40
40(39)	Megascleres oxeas; sponge soft; pale lavender in color; common .... *Haliclona permollis* [Oxeas only 100 × 3–7 μm]
	Megascleres tetractinal (calthrops); surface pattern lobulate; color pale yellow; rare but could be seasonally abundant .... *Plakina monolopha* [Triacts and tetraxons rays 20–30 × 3–4 μm; microxeas 70–200 × 7 μm; monolophotriaenes 26 μm]
41(39)	Color yellow gray to ocher; microscleres arcuate isochelae; rare .... *Xytopsiphum meganese* [Strongyles 210–240 × 3 μm; arcuate isochelae 16 μm]
	Color brown to black, mottled; microscleres birotulate-isochelae;[3] rare .... *Xytopsiphum kaneohe* [Wavy strongyles 200 × 4 μm; bipocillate isochelae 10–15 μm]
42(20)	Sponge having oscules elevated above sponge surface on large oscular projections or small cones 43
	Sponge lacking oscular elevations 52
43(42)	Microscleres absent 44
	Microscleres present 48
44(43)	Megascleres oxeas 45
	Megascleres styles; sponge massive with tapering oscular projections; color green, surface tuberculate; rare .... *Hymeniacidon chloris* [Styles 300–400 × 5–7 μm]
45(44)	Sponge with dermal membrane clearly separable from underlying tissues 46
	Sponge lacking any dermal specialization; sponge tubular; color blue violet; rare .... *Haliclona aquaeducta* [Oxeas only 120–150 × 5–7 μm]
46(45)	Color uniform throughout the sponge 47
	Color dark brown externally, dull yellow internally; surface minutely conulose; common on pilings and rafts .... *Halichondria melanadocia* [Oxeas 200–512 × 2–13 μm]
47(46)	Color yellow, texture papery, body semitransparent; rare .... *Pellina eusiphonia* [Oxeas 450–480 × 12–15 μm]
	Color delicate turquoise blue, oscular tubes tending to be fluted; common on pilings and rafts .... *Halichondria coerulea* [Oxeas 240–600 × 2–12 μm]

---

[3]Editor's note: De Laubenfels (1950, p. 12; 1951, p. 259) referred to these as U-shaped arcuate chelas with thick shafts and clads so small as to appear at first glance as sigmas.

48(43) Dermal membrane containing areas of more dense or contrasting pigment, subdermal cavities prominent; color variable—pink, lavender, yellow, orange, red, dull blue; extremely common .... *Mycale cecilia*
[Styles 240–250 × 4–6 μm; 30–42 μm; anisochelae 15–24 μm]
    Sponge lacking areas of dense or contrasting pigmented dermis 49

49(48) Surface with prominent exhalant channels; color dull orange, body often drawn out into long fingerlike projections which divide small cylindrical branches; common pilings .... *Tedania macrodactyla*
[Tylotes 165–210 × 2.5–4.0 μm; styles 150–200 × 3–7 μm; onychaetes 160–140 × up to 2.0 μm]
    Prominent exhalant channels lacking 50

50(49) Microscleres toxas; color vivid violet; sponge tubular, several tubes tend to be laterally fused; texture extremely soft; common .... *Toxadocia violacea*
[Oxeas 120–140 × 4–7 μm; toxas 60 × 1 μm]
    Microscleres not toxiform 51

51(50) Color blood red externally, orange red internally; microscleres onychaetes; surface slimy; growth from a thick basal mat bearing elevated oscular chimneys; common .... *Tedania ignis*
    Color bright orange externally, white internally; microscleres isochelae and sigmas; oscules elevated on fingerlike projections and closed by a membranous diaphragm; common .... *Damiriana hawaiiana*
[Tylotes 200 × 5 μm; oxeas 200 × 8 μm; arcuate isochelae 15–30 μm; sigmas 13 μm]

52(42) Color black 53
    Color not black 54

53(52) Sponge surface smooth and shiny; no mineral or spongin skeleton present; occasional .... *Chondrosia chucalla*
[No skeleton]
    Sponge surface velvety, consistency tough and elastic; microscleres birotulate; occasional .... *Iotrochota protea*
[Strongyles 140–205 × 3–6 μm; styles 135–180 × 7–10 μm; bipocilli 12–15 μm]

54(52) Spiny spicules (acanthostyles) represented among megascleres 55
    No spiny spicules among megascleres 56

55(54) Sponge erect, digitate and branching, surface smooth; subdermal spaces prominent; megascleres all monactinal, styles of two sizes and acanthostyles, microscleres palmate isochelae and toxas; color red; common .... *Clathria procera*
[Large styles 290–360 × 12–16 μm; dermal styles 120–160 × 4 μm; acanthostyles 50–105 × 5–6 μm; toxas 25–50 μm; palmate isochelae 16 μm]
    Sponge massive, surface lumpy, oscules large (6 mm) closed by thin membranes; color bright red; megascleres tornotes and acanthostyles, microscleres isochelae and sigmas; occasional on docks, pilings .... *Myxilla rosacea*

[Tornotes 160 × 3 μm; acanthostyles 140 × 8 μm; isochelae 15 μm; sigmas 18–30 μm]

56(55) Sponge cylindrical ramose to repent with high content of horny fiber; megascleres oxeas; color brown to lavender; common .... *Callyspongia diffusa*
[Oxeas only 70–108 × 4–10 μm]

Sponge massive, often produced into long lobate projections which are sometimes fused laterally; surface slightly wrinkled or tuberculate; external color variable—turquoise, rose red, green, orange red, or violet; internal and base color always yellow; megascleres tylostyles; very common but seasonal .... *Terpios zeteki*
[Tylostyles, two sizes—300–700 × 4–14 μm and 200 × 2 μm (often with 4 lobed heads)]

KEY TO CALCAREA KNOWN FROM HAWAIIAN WATERS

1 Color white 2
  Color yellow; sponge a mass of delicate anastomosing tubes ... *Clathrina* sp.

2(1) Sponge in the form of a hollow branched cylinder; spicules triacts, quadriradiates, and oxeas; color white.............. *Leuconia kaiana*
  Sponge amorphous; spicules triacts of two sizes; color white .... *Leucetta* sp.

LIST OF HAWAIIAN DEMOSPONGIAE[4]

Order DICTYOCERATIDA
Verongiidae
  Psammaplysilla purpurea (Carter 1880) [syn. *Hexadella pleochromata* de Laubenfels, 1950 according to Bergquist, 1967]

Spongiidae
  Spongia oceania de Laubenfels 1950 [also reported in de Laubenfels, 1951, 1957]

Dysideidae
  Dendrilla cactus (Selenka 1867) [first Hawaiian record; see de Laubenfels (1948) for morphology, figures]
  Dysidea herbacea (Keller 1889) [Bergquist, 1967]
  Dysidea sp. [as *Dysidea avara* (Schmidt) in de Laubenfels, 1950]

Order DENDROCERATIDA
Aplysillidae
  Aplysilla rosea (Barrois 1876) [Bergquist, 1967: encrusting, with markedly conulose surface and rose pink color; widespread on undersurfaces of rocks]
  Aplysilla sulphurea Schulze 1878 [Bergquist, 1967]
  Aplysilla violacea Lendenfeld 1883 [Bergquist, 1967]
  Pleraplysilla hyalina de Laubenfels 1950

---

[4]Species preceded by an asterisk have not been considered in the key. Information in brackets gives Hawaiian references, synonyms, or other details. Species included are from depths of 100 m or less.

Order HAPLOSCLERIDA
- **Haliclonidae**
  - *Haliclona aquaeducta* (Schmidt 1862) [as *Reniera aquaeducta* in de Laubenfels, 1951]
  - *Haliclona permollis* (Bowerbank 1866) [de Laubenfels, 1951]
- **Callyspongiidae**
  - *Callyspongia diffusa* (Ridley 1884) [de Laubenfels, 1950, 1951]
- **Desmacidonidae**
  - *Iotrochota protea* (de Laubenfels 1950) [as *Hiattrochrota protea*]
  - *Xytopsiphum kaneohe* de Laubenfels 1950
  - *Xytopsiphum meganese* de Laubenfels 1951
  - **Xytopsues zukerani* de Laubenfels 1957 [dredged, 75 m]

Order POECILOSCLERIDA
- **Adociidae**
  - *Pellina eusiphonia* Ridley 1884 [as *Pellina sitiens* in de Laubenfels, 1957]
  - *Petrosia puna* de Laubenfels 1951
  - *Toxadocia violacea* de Laubenfels 1950 [syn. *Kaneohea poni* and *Neoadocia mokuoloe* de Laubenfels, 1950, according to Bergquist, 1967]
- **Phorbasidae**
  - *Axociella kilauea* de Laubenfels 1951
  - *Damiriana hawaiiana* de Laubenfels 1950 [also reported in de Laubenfels, 1951]
- **Mycalidae**
  - *Mycale cecilia* de Laubenfels 1936 [in de Laubenfels, 1950; syn. *Mycale maunakea* de Laubenfels, 1951, *Mycale manuakea (sic)* according to Bergquist, 1967]
  - *Mycale (Carmia) contarenii* (Martens 1824) [as *Carmia contarenii* in de Laubenfels, 1951]
  - **Ulosa rhoda* de Laubenfels 1957 [dredged, 20-50 m]
  - *Zygomycale parishi* (Bowerbank 1875) [de Laubenfels, 1950, and Bergquist, 1967]
- **Tedaniidae**
  - *Tedania ignis* (Duchassaing and Michelotti 1864) [de Laubenfels, 1950, 1951, and Bergquist, 1967]
  - *Tedania macrodactyla* (Lamarck 1814) [Bergquist, 1967]
- **Microcionidae**
  - *Clathria procera* (Ridley 1884) [Bergquist, 1967]
  - *Microciona maunaloa* de Laubenfels 1951 [also in de Laubenfels, 1957]
  - *Microciona* sp.
- **Myxillidae**
  - *Hymedesmia* sp.
  - *Myxilla rosacea* (Lieberkühn 1859) [de Laubenfels, 1950, 1957]
  - *Naniupi ula* de Laubenfels 1950

Order HALICHONDRIDA
- **Halichondridae**
  - **Ciocalypta penicillus* Bowerbank 1864 [in Bergquist, 1967: offshore (60 m depth) sandy substrate, appears as a buried massive base from

which tapering fluted fistules arise]
 Halichondria coerulea Bergquist 1967
 Halichondria dura Lindgren 1897 [de Laubenfels, 1951]
 Halichondria melanadocia de Laubenfels 1936 [Bergquist, 1967]
 Rhaphisia myxa de Laubenfels 1951
Hymeniacidonidae
 *Densa distincta de Laubenfels 1957 [dredged, 50 m]
 Hymeniacidon chloris de Laubenfels 1950

Order AXINELLIDA
Raspailiidae
 Eurypon nigra Bergquist 1967 [as *Eurypon distincta* (Thiele) in de Laubenfels, 1957]

Order HADROMERIDA
Suberitidae
 Prosuberites oleteira de Laubenfels 1957
 Terpios granulosa Bergquist 1967
 Terpios zeteki (de Laubenfels 1936) [de Laubenfels, 1950, 1951, 1954a]
Spirastrellidae
 **Diplastrella spiniglobata* (Carter 1879) [Bergquist, 1967: very thin encrusting, uneven thickness, brick red alive]
 Spirastrella coccinea (Duchassaing and Michelotti 1864) [syn. *Spirastrella keaukaha* de Laubenfels, 1951, according to Bergquist, 1967]
 **Spirastrella vagabunda* Ridley 1884 [Bergquist, 1967: found on sandy bottom in Kaneohe Bay at depth of 10 m, body of sponge buried in sand and stout oscular projections protrude 5-6 cms above surface]
Clionidae
 Cliona vastifica Hancock 1849 [de Laubenfels, 1950]
Tethyidae
 Tethya cf. *diploderma* Schmidt 1870 [as *Tethya diploderma* in de Laubenfels, 1950, 1951, 1954a]
 Tethya sp.

Order CHORISTIDA
Stelletiidae
 Asteropus kaena (de Laubenfels 1957) [as *Stellettinopsis kaena;* also found in shallow water (Bergquist, pers. comm.)]
 Stelletta debilis (Thiele 1900) [as *Myriastra debilis* in de Laubenfels, 1951]
Jaspidae
 *Jaspis pleopora (de Laubenfels 1957) [as *Dorypleres pleopora,* dredged, 50 m]
 Zaplethea digonoxea de Laubenfels 1950
Geodiidae
 *Erylus proximus Dendy 1916 [de Laubenfels, 1957, dredged (50 m) and shallow water]
 Geodia gibberella de Laubenfels 1951
Chondrosiidae
 Chondrosia chucalla de Laubenfels 1936 [de Laubenfels, 1951]

Order HOMOSCLEROPHORIDA
Plakinidae
Plakina monolopha Schultz 1880 [de Laubenfels, 1951]
Plakortis simplex Schultz 1880 [de Laubenfels, 1950, 1951]
Order LITHISTIDA
Leiodermatium sp.

LIST OF HAWAIIAN CALCAREA

Grantiidae
Leuconia kaiana de Laubenfels 1951
Leucettidae
Leucetta sp. [as *Leucetta solida* in de Laubenfels, 1950, 1951, 1957]
Leucosoleniidae
Clathrina sp. [as *Leucosolenia eleanor* in de Laubenfels, 1954a]

ADDITIONAL SPONGES KNOWN FROM HAWAIIAN WATERS[5]

Class DEMOSPONGIAE

HAPLOSCLERIDA
Haliclona flabellodigitata Burton (3)
POECILOSCLERIDA
Adocia gellindra de Laubenfels (2)
Lissodendoryx calypta de Laubenfels (3)
Microciona haematodes de Laubenfels (1)
AXINELLIDA
Axinella solenoides de Laubenfels (1)
Homaxinella anamesa de Laubenfels (1)
Phycopsis aculeata Wilson (3)
Axechina lissa de Laubenfels (1) [dredged, depth not given]
Eurypon distincta (Thiele) (3)
HADROMERIDA
Anthosigmella valentis de Laubenfels (1) [=*Spirastrella valentis*]
Timea xena de Laubenfels (2)
Kotimea tethya de Laubenfels (2, 3)
CHORISTIDA
Erylus rotundus Lendenfeld (3)
HOMOSCLEROPHORIDA
Oscarella tenuis Hentschel (2)

Class CALCAREA

LEUCOSOLENIDA
Leucosolenia eleanor Urban (2)
Leucosolenia vesicula (Haeckel) (3)
Leucetta solida (Schmidt) (3)

---

[5]The sponges listed here are those (1) known only from depths greater than 100 m (see de Laubenfels, 1957), (2) recorded only from the Waikiki Aquarium (see de Laubenfels, 1954a), or (3) considered doubtfully identified (see de Laubenfels, 1951, 1957).

GLOSSARY (PORIFERA)

acantho-: prefix meaning spiny, as in acanthostyle (Pl. 1, Fig. 1).

actine: a ray or point; used as a suffix when referring to the number of rays or points of a sponge spicule (for example, **monactines, diactines, hexactines, triactines**).

anatriaene: see under **triaene**.

anisochela: chela with unequal ends (Pl. 1, Fig. 2).

asters: microscleres which are starlike in shape.

birotulate: a sponge spicule having two wheel-shaped cups at the ends (Pl. 1, Fig. 3).

budding: asexual reproduction by which portions (buds) of a sponge colony may break off and grow into new colonies.

calthrops: tetraxonal megasclere having the four rays equal or nearly equal in dimensions.

centrotylote oxea: spicule pointed at both ends with an expansion in the middle (Pl. 1, Fig. 4).

chela: microsclere type resembling a pair of anchor flukes (Pl. 1, Fig. 5).

conules: cone-shaped elevations of the sponge surface usually supported by a "primary" spongin fiber or a tract of spicules (Pl. 1, Fig. 8).

cortex: outer layer of sponge clearly set off from the internal layer; where this division is definite, the term **cortex** is used in preference to **ectosome**.

dermis: or **dermal membrane** is the bounding layer of the sponge; this may contain spicules and is pierced by pores and oscules (Pl. 1, Fig. 8).

desmas: much branched and irregular spicule of underlying monaxon, triaxon, or tetraxon pattern; desmas are usually united into a network and make up the skeleton of lithistid sponges, for example, *Leiodermatium* (Pl. 1, Fig. 10).

diactine: spicule formed by growth in two directions along a straight or curved axis and hence similar at each end, for example, **oxea, tylote**.

ectosome: immediate subdermal layer of the sponge where this has a spicule complement distinct from that of the deeper layers but where there is little histological differentiation.

endosome: internal mass of the sponge body.

euasters: microscleres having several equal rays radiating from a central point; ends of rays vary from pointed (oxeote) to rounded (strongylote) (Pl. 1, Fig. 9).

gemmule: asexual reproductive units consisting of a mass of food-filled cells surrounded by a heavy protective coat reinforced with spicules.

isochela: chela with both ends similar (Pl. 1, Fig. 5).

lipostomous: condition, common in encrusting sponges, in which neither pores nor oscules are visible.

megascleres: large spicule types which are the major structural elements of the sponge (Pl. 1, Figs. 11, 13, 22, 23; Pl. 2, Fig. 8).

microscleres: relatively small, cortical or ectosomal spicules, often of peculiar form (Pl. 1, Figs. 2, 3, 4, 8, 9, 11, 12, 13; Pl. 2, Figs. 2, 5, 6).

monolophotriaene: modified microcalthrops where one ray divides to produce three terminal branches (Pl. 1, Fig. 7).

onychaetes: fine microscleres which are diactinal, often relatively long, and which differ from raphides in being microspined; characteristic of the genus *Tedania* (Pl. 1, Fig. 15).

oscules: exhalant openings for the water current.

ostia: see **pores**.

oxea: diactinal megasclere pointed at both ends (Pl. 1, Fig. 11).

oxyaster: asterose microsclere with sharp pointed rays and a small centrum (Pl. 1, Fig. 12).

plagiotriaene: megasclere with four actines, one of which (the rhabd) is longer than the other three (clads); the clads being directed slightly forward from the point of intersection (Pl. 1, Fig. 23).

pores: or **ostia** are very small incurrent openings.

quadriradiate: tetraxonal megasclere with four rays more or less in the same plane or with one ray bent; especially common in some calcareous sponges (Pl. 2, Fig. 1); see under **tetraxon** (b).

raphide: fine oxeote microsclere (Pl. 1, Fig. 24).

sigma: microsclere of C or S shape (Pl. 2, Fig. 2).

spheraster: asterose microsclere with a definite centrum and many rays, usually pointed (Pl. 2, Fig. 5).

spiraster: vermiform or rodlike microsclere with spines often arranged in sigmoid fashion around the axis (Pl. 2, Fig. 6).

spongin: collagenous (proteinaceous) material important in skeleton formation of most sponges: Spongin, a halogenated scleroprotein, occurs as fibers or cementing substance at angles between spicules.

True collagen is more universal, occurring as long unbranched fibrils in the matrix (Pl. 1, Fig. 8; Pl. 2, Fig. 4).
sterraster: modified asterose microsclere in which the rays are reduced to small projections from the spherical surface (Pl. 2, Fig. 7).
strongyle: diactinal megasclere in which both ends are rounded (Pl. 1, Fig. 25).
strongyloxea: megasclere in which the neck narrows at one end and then usually expands slightly, becoming rounded, and at the other end it is pointed or at least tapered; characteristic of the genus *Tethya* (Pl. 2, Figs. 3a, 3b).
style: monactinal megasclere in which one end is rounded and the other pointed (Pl. 1, Fig. 14).
subtylostyle: monactinal megasclere with one end pointed, the other rounded, but slightly expanded.
tetract: see **tetraxon**.
tetraxon: megasclere with four rays: (a) each ray pointed in a different plane and direction (for example, **calthrops, triaenes);** (b) the rays generally in the same plane (for example, **quadriradiates).**
tornote: diactinal spicule, sharply tapered to points at each end (Pl. 1, Fig. 13).
toxa: bow-shaped microsclere (Pl. 1, Fig. 16).
triact: megasclere with three rays in approximately the same plane but at different angles (Pl. 1, Fig. 17); sagittal triacts have two angles equal and one angle different (Pl. 1, Fig. 18); especially common in some calcareous sponges; see also under **triaxon** (b).
triaene: tetractinal megasclere with one ray greatly enlarged, the subsidiary rays (clads) may be straight **(orthotriaene),** projected slightly forward **(plagiotriaene),** projected markedly forward **(protriaene),** curved back **(anatriaene),** or branched **(dichotriaene)** (Pl. 1, Figs. 22, 23).
triaxon: megasclere with 3 axes: (a) if these cross at right angles and thus produce six actines or rays, this is a hexactine condition characteristic of the Class Hexactinellida; (b) if the axes intersect and each has only a single ray this is the triactine (triact) condition which is common in the Class Calcarea.
tylaster: small aster with no centrum and having tylote (knobbed) ends to the rays (Pl. 1, Fig. 19).
tylospheraster: small aster with a centrum and tylote (knobbed) ends on the rays (Pl. 1, Fig. 20).
tylostyle: monactinal megasclere with a knob at one end and pointed at the other (Pl. 2, Fig. 8).
tylote: diactinal megasclere with rounded or oval knobs at each end (Pl. 1, Fig. 21).

REFERENCES (PORIFERA)

Bergquist, P. R.
    1965. Sponges of Micronesia, Part I. The Palau Archipelago. *Pacific Science* 19(2): 123-204.
    1967. Additions to the Sponge Fauna of the Hawaiian Islands. *Micronesica* 3(2): 159-173. 8 figs, 1 pl.
Edmondson, C. H.
    1946. Reproduction in *Donatia deformis* (Thiele). *B. P. Bishop Mus. Occ. Pap.* 18(18): 271-282.
Haeckel, E.
    1872. *Die Kalkschwamme: Eine Monographie System der Kalkschwamme.* Vol. 2. 148 pp.
Laubenfels, M. W. de
    1948. *The Order Keratosa of the Phylum Porifera: A Monographic Study.* Allan Hancock Foundation Publ., Occ. Pap. 3. 217 pp.
    1950. The Sponges of Kaneohe Bay, Oahu. *Pacific Science* 4(1): 3-36.
    1951. The Sponges of the Island of Hawaii. *Pacific Science* 5(3): 256-271.
    1954a. Occurrence of Sponges in an Aquarium. *Pacific Science* 8(3): 337-340.
    1954b. *The Sponges of the West Central Pacific.* Oregon State Monogr. Zoology 7. 306 pp.
    1957. New Species and Records of Hawaiian Sponges. *Pacific Science* 11(2):236-251.
Lendenfeld, R. von
    1910. The Sponges, 2. The Erylidae. *Mus. Comparative Zoology Mem.* 41(2): 267-323. Harvard Univ.
Munsell, A.
    1942. *Book of Color.* (Pocket ed.) 2 vols. Baltimore: Munsell Color Co.

CNIDARIA (COELENTERATA)

THE CNIDARIANS or coelenterates are organisms belonging to the most primitive phylum having a definite form, symmetry, and true organs. All forms are basically radially symmetrical and possess digestive, muscular, nervous, and sensory systems. Diagnostic of this phylum are nematocysts, stinging organoids containing an eversible hollow tube which, when stimulated, can inject a toxin, be used as an adhesive, or carry out other similar functions. Two basic body forms are found—the polyp and the medusa. The polyp is an attached, asexually reproducing, cylinderlike form with a mouth and tentacles at the oral end. The free-swimming, solitary, sexually reproducing medusa is reduced in height and expanded radially. Both forms occur in the life cycle of most individuals, except among the Anthozoa. Alternation of generations (metagenesis, or the occurrence of both sexual and asexual phases during the life cycle) is common among many cnidarians.

A central digestive cavity is surrounded by the body wall. This is composed of outer epidermis and inner gastrodermis separated by an intermediate layer (mesoglea). The digestive system includes a mouth, pharynx (gullet), and coelenteron (gastrovascular cavity). The nervous sytem is a network of nerve cells; no definite brain exists. A skeleton may or may not be produced. When present, it can be either separate spicules or a more or less solid exoskeleton or endoskeleton composed of calcareous, horny, or chitinous substances.

Sexual reproduction results in a free-swimming larva (planula) which usually is capable of settling on a suitable substrate and growing into a mature polyp; in some groups it may develop more or less directly into a medusa.

The phylum is divided into three classes—Hydrozoa, Scyphozoa, and Anthozoa.

Class HYDROZOA

The Hydrozoa are characterized by the development of both polypoid and medusoid forms and the species are generally considered to have an "alternation of generations" between the two. However, in many species the medusoid form is suppressed, either remaining attached to the asexually reproductive polyp or even reduced to gonadal masses on the parent polyp. On the other hand, some medusae never have an attached polypoid generation. The polypoid form usually grows by asexual reproduction into a branching colony, called a hydroid colony; in these the individual polyps may be naked or athecate, or surrounded by a cup of exoskeleton, the thecate hydroids. The medusae found in this class can be distinguished from those of the class Scyphozoa by the presence of an inner shelf, called a velum, around the aperture of the bell. Within the class the coelenteron is not divided by septa.

Three orders (Hydroida, Siphonophora, and Chondrophora) are verified by recent collections in Hawaiian waters. Specimens of the coral-like order Stylasterina labeled as being from Hawaii in the Museum National d'Histoire Naturelle, Paris, were considered to have an erroneously credited locality (Boschma, 1959). Likewise, no species of "fire" corals (order Milleporina) are known in Hawaiian waters.

REFERENCE (HYDROZOA)

Boschma, H.
    1959. Revision of the Indo-Pacific Species of the Genus *Distochopora*. *Bijdragen tot de Dierkunde* 29:121-171.

Order HYDROIDA

WILLIAM JOHNSON COOKE
University of Hawaii

Although the *Albatross* expedition reported 49 species of hydroids from the deeper waters around the Hawaiian Islands (Nutting, 1905), the shallow water hydrozoans have been little investigated. Edmondson (1946) identified only one form specifically—*Pennaria tiarella* McCrady—and listed three other families—Campanulariidae, Sertulariidae, and Plumulariidae—as having local representatives. Edmondson (1930) described four new species of creeping hydromedusae for the genus *Eleutheria*. Boone (1938) described as a new species *"Corydendrium splendidum"* collected from Kaneohe Bay by the *Alva* expedition. A single species (*Plumularia (Halopteris) buski* Bale) very similar to *Halopteris diaphana* (Heller) was recorded from Laysan by Hartlaub (1901).

The Hawaiian hydroid fauna is, however, a good deal richer than this. At least twenty-eight species of hydroid polyps or benthic hydromedusae are now reported in the shallow water of the main islands. Twenty-three species are identified with reasonable certainty, but the remainder are identified only to the generic level, or in some cases only to family because the specimens lack the distinguishing reproductive structures.

As one might expect, the shallow-water hydroid fauna of Hawaii is marked by a preponderance of cosmopolitan species. This is reasonable when one considers the distance from source areas. It should be noted, however, that many of the forms reported were collected from artificial habitats such as floats, pilings, and so forth, or from disturbed areas such as Kaneohe Bay. This would also tend to favor the establishment of colonizing (in the ecological, not morphological sense) species, which in general are quite widely distributed.

Coral reefs do not support many large and dramatic forms such as are found in temperate waters, and, while there are exceptions such as *Lytocarpus philippinus*, most of the coral reef forms from shallow water are small cryptic species.

Chu and Cutress (1954) list a hydromedusa of the genus *Liriope* (Geryoniidae) as causing dermatitis, but this is evidently not a common inhabitant in reef waters, because none have been collected recently. Kramp (1961, p. 238) considers that there is only one species in the genus *Liriope*, *L. tetraphylla* (Chamisso and Eysenhardt 1821).

Finally, while this report adds considerably to the number of hydroid species known from Hawaii, it is certain that more remain to be discovered, especially gymnoblastic forms, and those such as small campanularids. Any specimen not identifiable through the key and descriptions has probably not been collected here before. The only recourse for these is a search through the references in the bibliography.

KEY TO THE HYDROID FAMILIES KNOWN FROM HAWAIIAN WATERS

1   Polyps completely naked or with perisarc covering only to base of
        polyps [athecate hydroids—Suborder GYMNOBLASTEA] 2
    Polyps covered with perisarc when retracted [thecate hydroids—
        Suborder CALYPTOBLASTEA] 10

2(1) Polyps without perisarc; tentacles with discrete rings of nematocysts,
        that is, moniliform .... **Moerisiidae**
    Polyps and tentacles otherwise 3

3(2) Polyps with capitate or with capitate and filiform tentacles 4
    Polyps with only filiform tentacles 8

4(3) Only capitate tentacles present 5
    Both capitate and filiform tentacles present 7

5(4) Scattered capitate tentacles present 6
    Four oral capitate tentacles only .... **Cladonematidae** (young polyp)
                                                            or **Eleutheriidae**

6(5) Colonies large, gorgonianlike, with an internal perisarc of anastomosing
        chitinous or horny fibers .... **Solanderiidae**
    Small colonies with typical perisarc .... **Corynidae**

7(4) Polyps with scattered capitate tentacles distally, and a basal ring
        of filiform tentacles .... **Halocordylidae**
    Polyps with four oral capitate tentacles and four short, stiff basal
        filiform tentacles .... **Cladonematidae** (mature polyp)

8(3) Filiform tentacles scattered over the polyp .... **Clavidae**
    Filiform tentacles in basal ring only 9

9(8) Body of polyp not sharply differentiated, hypostome
        conical .... **Bougainvilliidae**
    Body of polyp sharply differentiated into a trumpet or bulb-shaped
        hypostome .... **Eudendriidae**

10(1) Hydrothecae placed on pedicels from the stem, or arising directly
        from stolons 11
    Hydrothecae without pedicels, usually adnate 12

11(10) Hydrothecae cup-shaped, placed on pedicels, never
        operculate .... **Campanulariidae**
    Tubular hydrothecae arising singly, operculum cone-shaped,
        composed of many flaps .... **Campanulinidae**

12(10) Tubular, adnate, incompletely delimited hydrothecae present, not
        capable of covering retracted polyp .... **Haleciidae**
    Completely delimited hydrothecae present, capable of covering
        retracted polyp 13

13(12) Hydrotheca with operculum .... **Sertulariidae**
    Hydrotheca without operculum 14

14(13) Nematophores absent; hydrothecae tubular .... **Syntheciidae**
Nematophores present; hydrotheca variously shaped .... **Plumulariidae**

Suborder GYMNOBLASTEA

Family **Moerisiidae**

The hydroids of this family are characterized by the presence of scattered, hollow, moniliform tentacles. Moerisid medusae have similar moniliform tentacles and bear the gonads on the stomach or its radial lobes (see Rees, 1958, for a fuller discussion of this family). Most representatives are found in brackish water environments.

Ostroumovia horii (Uchida and Uchida 1929). Specimens of a hydroid identical to that discussed by Uchida and Nagao (1959) were collected from a brackish pond in a lava flow near Honokohau on the island of Hawaii. The polyp is as long as 2 cm, with most of the length a highly contractile stalk (Fig. 1). The hydranth itself is 1 mm to 2 mm long with 4 to 15 tentacles (the number increasing with age) and a well defined hypostome. The tentacles bear the nematocyst rings characteristic of the group and are also slightly capitate. The nematocyst rings are not quite complete, failing to fully encircle the tentacle, although this can be confirmed only in extended live individuals.[1] Scattered among the tentacles of older individuals are medusae buds, as many as 10 in number, in all stages of development. The most complete medusae appear close to release. These are about 0.5 mm tall and 0.4 mm in diameter and bear four moniliform tentacles. Asexual budding occurs with new hydranths being formed from the lower part of the hydranth and stalk. No basal perisarc or chitinous disk was observed, though all the older preserved specimens had been removed from the substrate, and all the live individuals were young. Kramp (1961) considered the species as a *Moerisia*. The recording of this species from both Japan and Hawaii is quite interesting, since most species in this family have a severely restricted distribution, unless transported by human activity.

Family **Solanderiidae**

This family is characterized by the presence of a perisarc composed of anastomosing chitinous fibers and polyps with scattered capitate tentacles. Since there is an external layer of ectoderm on the outer surface of this perisarc, it was at one time assumed to be a mesogleal skeleton, as in the gorgonians, to which the colonies bear a superficial resemblance. However, Vervoort (1966) demonstrated that the perisarc is indeed an ectodermal product, every fiber in the network being completely surrounded by ectoderm, and the endoderm being limited to tubes or strands through the ectoderm. Reproduction is always by means of fixed gonophores. The family includes encrusting forms and upright branching forms. There are apparently several species found in the shallow water.

Solanderia (?) *minima* (Hickson 1903). The colonies are up to 15 cm tall, with a thick main stem, and abundant branches mostly in the same plane, so that the colonies are distinctly flabellate. The stem and branches are composed of a mass of anastomosing fibers. The locations of the polyps are marked by various elaborations of the perisarc. On the thin distal branches, the polyps are flanked by two

---

[1]The nematocysts are as described by Uchida and Nagao (1959) in kinds and sizes.

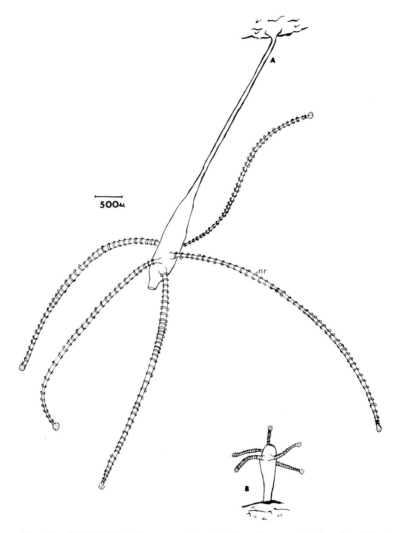

Figure 1.—HYDROIDA. *Ostroumovia horii*. A view of two small polyps from life, showing extended and contracted specimens; *nr*, nematocyst rings.

spurlike projections, whereas on the stem and thicker branches, there is usually a low ring or proximal ridge of coalesced spines. The color of the colonies is dark brown, tending to a lighter yellow brown near the tips of the branches, where the fibers are not so densely packed.

According to Vervoort's (1962) key, this species would be assigned to *S. minima*, originally described from Zanzibar, although it possesses features characteristic of *S. secunda*, a form known from the central Pacific. (A local specimen from 120 meters can definitely be assigned to *S. secunda*, while Nutting (1905) listed *S. fusca* (as *Ceratella fusca*) from deeper Hawaiian waters.) Unfortunately, no polyps were well enough preserved on this material to allow critical examination, and the colony was sterile. The description is based on colonies collected by Richard Grigg near Kipapa Island in Kaneohe Bay, Oahu, from an underwater cave at a depth of 3 m to 4 m.

Solanderia misakinensis (Inaba 1892). A colony identified as this species by W. Vervoort (Rijksmuseum van Natuurlijke Historie, Leiden) was collected under a ledge off the Halona Blowhole, Oahu, at a depth of about 7 m. The polyps are small, and male and female gonophores occur on separate colonies (Figs. 2, 3). This species differs from *S. minima* in lacking the hydrophore completely. Previous records of *S. misakinensis* are exclusively from Japan.

Family **Corynidae**

This family is characterized by the hydranths having scattered capitate tentacles. Reproduction may either be through free medusae (in genera such as *Sarsia*) or through sporosacs (as in *Coryne*). Specimens of an unidentified form have been collected from burrows excavated by the alpheid shrimp, *Alpheus deuteropus* Hilgendorf, in the coral *Porites lobata*. The polyps are small, 0.5 mm in height, with at most 12 scattered tentacles (Fig. 4). The polyps are light brown and either arise singly or form small, irregularly branched colonies. This may be the same hydroid (*Sarsia (Syncoryne) mirabilis* (Agassiz 1862)) mentioned by Chu and Cutress (1955)

Figure 2.—HYDROIDA. *Solanderia misakinensis*. Entire colony from Hawaiian Islands (Rijksmuseum van Natuurlijke Historie, Leiden, photo).

Figure 3.—HYDROIDA. *Solanderia misakinensis*. Close-up of branches showing gonophores and polyps (Rijksmuseum van Natuurlijke Historie, Leiden, photo).

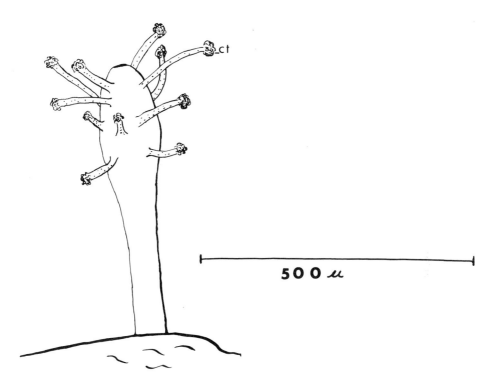

Figure 4.—HYDROIDA. A polyp of the Family Corynidae (*Sarsia mirabilis* ?) collected from *Porites lobata*; *ct*, capitate tentacles.

as the cause of a dermatitis, although all specimens examined have been sterile and precise identification has not been possible.

Family **Cladonematidae**

The description of this family is based on the medusa, the larger and more conspicuous phase. Both the polyp and the benthic medusae are likely to be encountered in shallow waters. The medusae are bell-shaped creeping forms with a well-formed manubrium bearing oral tentacles. The hydroids are small forms with an oral whorl of four capitate tentacles and usually with four short basal filiform tentacles.

Cladonema radiatum Dujardin 1843. The bell of the medusa is thin-walled and transparent, being somewhat higher than broad, about 3 mm high (Fig. 5a). A small hump is present on the top center over the manubrium, which is dark brown in color and lined with white vertical stripes. The manubrium is also ridged and bears six capitate oral tentacles. There are six radial canals arising from the center, three of which immediately bifurcate, so that there are nine radial canals on the bell. There are nine marginal tentacles, each of which has an ocellus at its base. The proximal parts of these tentacles are swollen and orange and bear up to ten peduncles, which can be used for attachment to the substrate. The main arm of the tentacle will bear up to ten side branches with nematocyst clusters. The main arms may extend to much more than twice the height of the bell. The hydroid stage consists

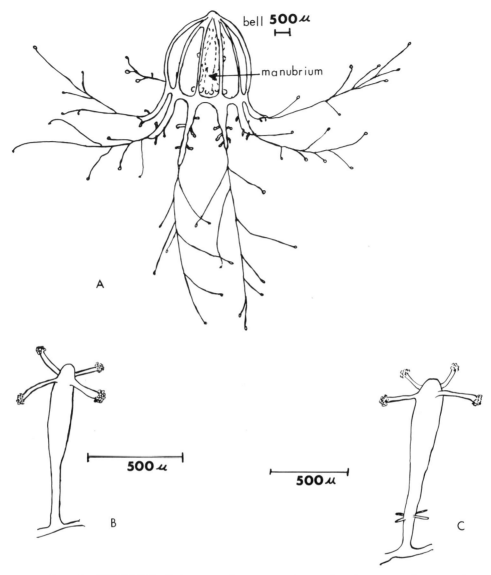

Figure 5.—HYDROIDA. *Cladonema radiatum: a,* a mature medusa; *b,* a young polyp; *c,* a mature polyp, showing the four basal tentacles.

of polyps approximately 1 mm high which arise singly from a creeping stolon (Fig. 5*b*). These polyps have four capitate tentacles arranged at right angles just below the mouth and four short, stiff filiform tentacles lower on the body (Fig. 5*c*). The medusae are often found on the alga *Ulva* in relatively quiet waters. This is a cosmopolitan species, with a good deal of variation between forms from different areas.

Family Eleutheriidae

This family is characterized by small creeping medusae which have a ring of nematocysts around the edge of the much flattened bell. There are no oral tentacles, and the marginal tentacles usually have peduncles, and are branched. While Ed-

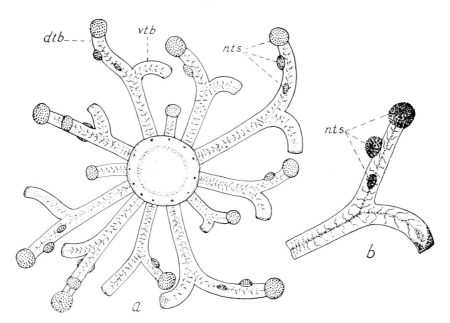

Figure 6.—HYDROIDA. *Staurocladia bilateralis: a,* dorsal view with tentacles extended; *b,* tentacle showing clusters of nematocysts; *dtb,* dorsal tentacular branch; *nts,* nematocysts; *vtb,* ventral tentacular branch (from Edmondson, 1930).

mondson (1930) described the local forms as *Eleutheria,* they were considered in the genus *Staurocladia* by Kramp and other workers. Kramp (1952) lists a *Staurocladia* from the southern coast of Chile as identical to *S. oahuensis*. There is a good deal of variability in the characteristics used to delimit the species, and it is with some hesitation that the forms seen have been assigned to the two species listed below. Individuals fitting the descriptions of the other two species, *S. acuminata* (Edmondson) and *S. alternata* (Edmondson) are likely to exist, but I have not seen them. In addition to the medusae reproducing by asexual fission, the life cycle includes small polyps (very similar to those of *Cladonema*) with four oral tentacles.

(?) *Staurocladia bilateralis* (Edmondson 1930) [syn. *Eleutheria bilateralis* Edmondson]. This small creeping medusa is less than 1 mm in bell diameter, or 1.5 mm in total diameter. The specimens observed had ten tentacles with nematocyst clusters on the sides of the tentacles as well as on the dorsal surface (Fig. 6). They conform in general to Edmondson's description of *E. bilateralis*. The specimens were collected from *Ulva* at Black Point, Oahu.

(?) *Staurocladia oahuensis* (Edmondson 1930) [syn. *Eleutheria oahuensis* Edmondson]. This form is somewhat smaller, perhaps 1 mm or so in total diameter. Most individuals seen had only one dorsal nematocyst cluster on the tentacles, although sometimes two were observed (Fig. 7). Thus, it is most like Edmondson's description of *E. oahuensis*. These were also collected from the same locality as the previous forms.

Family **Halocordylidae**

This family is characterized by having a basal ring of filiform tentacles and scattered capitate tentacles on the upper body of the polyp. The colonies are regularly branched and are covered with a thick perisarc which extends to the base of

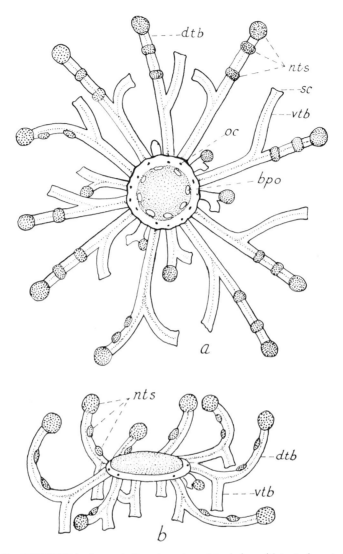

Figure 7.—HYDROIDA. *Staurocladia oahuensis: a,* dorsal view with tentacles extended; *b,* lateral view, animal creeping; *bpo,* opening of brood pouch; *dtb,* dorsal tentacular branch; *nts,* nematocysts; *oc,* ocellus; *sc,* sucker; *vtb,* ventral tentacular branch (from Edmondson, 1930).

polyps. Reproduction is by means of rudimentary medusae which may liberate gametes before or after being themselves released. The name *Halocordyle* for the only genus in this family was adopted by modern workers over the name *Pennaria,* a genus which did not include the present species.

Halocordyle disticha (Goldfuss 1820) [syn. *Pennaria tiarella* McCrady of Edmondson, 1946; *Corydendrium splendidum* Boone 1938]. The colonies are large, as tall as 30 cm, with a dark brown to black perisarc which is usually overgrown with diatoms and algae. Branching is alternate, with the polyps borne on the upper surfaces of the branches. The polyp has a circle of 10 to 18 filiform tentacles at the base and as many as 12 capitate tentacles on the upper part of the hydranth (Fig. 8). The hypostome is large and highly mobile. The body of the polyp is white with a reddish

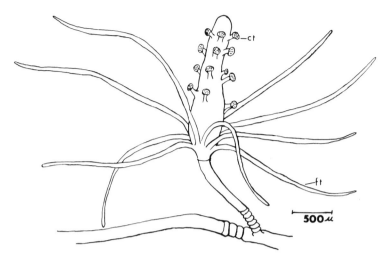

Figure 8.—HYDROIDA. *Halocordyle disticha*, a mature polyp: *ct*, capitate tentacles; *ft*, filiform tentacles.

tinge. Annulations occur on the branches which bear the polyps, sometimes completely covering these branches. The perisarc on the main stem and side branches is also annulated at various intervals. The medusae are often a brighter red. They arise from the region of the filiform tentacles. Previous records of this species in Hawaii include *Pennaria tiarella* which Vervoort (1968) considers, together with other synonyms, to be the same as the classical European form *Halocordyle disticha*. Species differentiations have been based on considerations of the number and orderliness of the capitate tentacles, the pattern of annulation, and other criteria which can show great intraspecific or intracolonial variation. It seems highly likely that all these species are forms of a worldwide species, which would be *H. disticha*. Boone's (1938) *Corydendrium splendidum* is synonymized with *H. disticha*, since the figure she gives for this species is clearly a poorly preserved *Halocordyle*. The specimen she examined was collected from Kaneohe Bay on Oahu, where *H. disticha* is abundant. Elsewhere on Oahu, it has been found at the Ala Wai Yacht Harbor, Kewalo Basin, Honolulu Harbor, and Keehi Marina, where it often grows attached to pilings, submerged lines, and boat bottoms. This species is known to sting when handled.

Family **Clavidae**

This family is characterized by the presence of only scattered filiform tentacles on the polyps. Colonies may consist of creeping hydrorhizas which bear individual polyps, or they may be upright branched forms. The colonies may reproduce either by means of fixed gonophores which liberate planulae or by means of free medusae. The growth pattern of the colony and the method of reproduction are the diagnostic features of the genera of this family. At least three members of this family are present in Hawaii; *Cordylophora caspia* (Pallas), *Turritopsis nutricula* McCrady, and a *Rhizogeton* sp.

Cordylophora caspia (Pallas 1771). This rather large hydroid is characterized by irregular loosely branched stems up to 15 mm tall and 0.2 to 0.25 mm in diameter with polyps given off semialternately and terminally (Fig. 9). The polyps are

1 mm to 2 mm in total length with 10 to 20 scattered filiform tentacles and are dull red distally. The perisarc of the stem ends well beneath the body of the polyp. Many gonophores are present on the colony, two to four (no more) on the stem, usually 1 mm or so below the terminal polyps (not as low on the side polyps). The orange gonophores are covered by the perisarc and are elongate-oval in shape, 1 mm long (including pedicel) by 0.3 mm maximum diameter. This is an extremely cosmopolitan hydroid (often recorded as *C. lacustris*) and has been found to be tolerant of salinities from 0 to 30‰. The Hawaiian material was collected by Diana Wong and John Maciolek from a brackish water (18 to 30‰) pond at Cape Kinau, Maui, where it was growing on the alga *Caulerpa serrulata*.

Turritopsis nutricula McCrady 1856. The colonies are small, usually not higher than 1.5 cm, and may consist of just a basal mat of stolons which give rise to individual hydranths, although higher colonies with irregular branchings are seen. The polyps are small, about 0.5 mm in height, white and pink, with approximately 20 scattered, highly contractile filiform tentacles (Fig. 10). The hydranth is spindle-shaped with a slender stalk. The perisarc is thin and brownish and ends just below the hydranth. Reproduction is by means of free medusae which are characteristically budded off in pairs just below the hydranths. The medusae are less than 1 mm

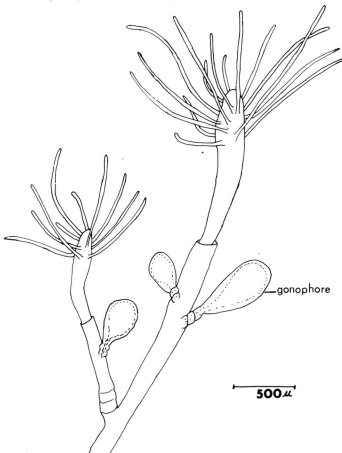

Figure 9.—HYDROIDA. *Cordylophora caspia*, two polyps with gonophores.

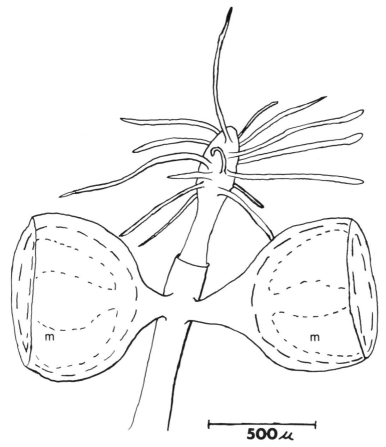

Figure 10.—HYDROIDA. *Turritopsis nutricula*, a polyp with two medusae buds.

in diameter with only a few tentacles when released, but grow to 3 mm or so and add many tentacles. It is likely that the medusa figured by Edmondson (1946; see Fig. 1b) represents a maturing *T. nutricula* medusa. This species was first described from the east coast of the United States, and has since been found to have a worldwide warm-water distribution. On Oahu, it is found, often growing on *Halocordyle* stems, in harbors such as Ala Wai Yacht Harbor.

(?) *Rhizogeton* sp. This hydroid has often been collected from alpheid crevices in the coral *Porites lobata*. The polyps are fair-sized, about 1 mm to 1.5 mm long, brownish red in life with white spots at the bases of the tentacles. There are 10 to 15 scattered filiform tentacles on the upper two-thirds of the hydranth (Fig. 11). The polyps are connected by an inconspicuous hydrorhiza which is very difficult to observe beneath the algal growth on the coral matrix. The disposition of the polyps and their size and structure indicate that these specimens may be assigned to the genus *Rhizogeton*, but no further identification is possible without fertile colonies.

Family **Bougainvilliidae**

This family is distinguished by the polyps, which have a basal whorl of filiform tentacles and an undifferentiated hypostome. Reproduction may be by means

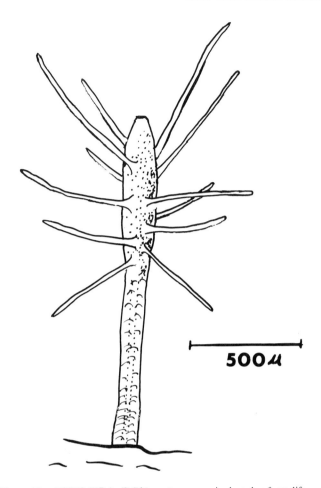

Figure 11.—HYDROIDA. (?)*Rhizogeton* sp., a single polyp from life.

of either fixed gonophores or free medusae, depending on the genus. At least two species are present in Hawaii.

Garveia sp. This hydroid often occurs on *Halocordyle* stems, some colonies being 6 mm to 8 mm high but most usually much smaller, with a series of individual hydranths on stems arising singly from the hydrorhiza. The light tan perisarc comes just to the base of the tentacles and is usually thickly encrusted with fine detritus. The hydranths are approximately 0.5 mm in length and 0.2 mm in diameter and have 10 to 14 tentacles. The gonophores, usually one to four in number, are borne just below the base of the hydranths (Fig. 12). They are pear-shaped, borne on short stems, and approximately 0.10 mm to 0.35 mm long, including the stems. The larger gonophores are probably female. This genus is widely distributed. The colonies studied were collected from *Halocordyle* colonies at the Ala Wai Yacht Harbor.

Bougainvillia ramosa (van Beneden 1844). This hydroid forms colonies which may reach a height of 7 cm to 8 cm, though the majority reach approximately 4 cm. Colonies consist of a straight stem, which may be fasicled in its lower portions and may have randomly arranged branches (Fig. 13). The hydranths are borne on individual pedicels 0.5 mm long alternatively arranged on the stem and branches every 1 mm to 2 mm. These pedicels are about 0.1 mm in diameter and are often annulated. The hydranths are about 0.5 mm long and 0.2 mm in diameter and bear 10 to 15 tentacles. The lower part of the hydranths may be covered with a very thin perisarc forming a pseudohydrotheca. The color of the polyps is white with a reddish tinge. One (sometimes two) medusa buds are borne just below the base of each polyp. These release medusae which are initially 0.5 mm in diameter, with four orange marginal tentacle clusters each having two tentacles, and with four oral tentacles at the end of a short manubrium. As the medusa matures, it may reach 2

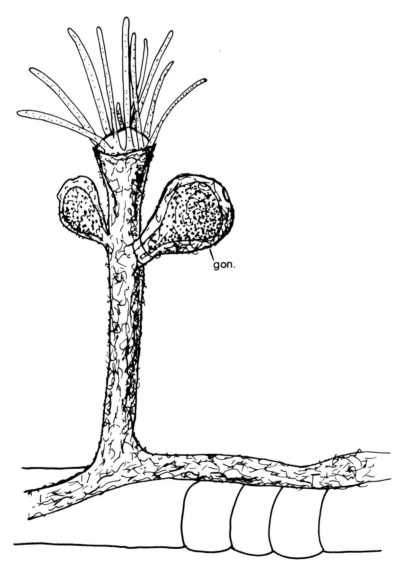

Figure 12.—HYDROIDA. *Garveia* sp., a single polyp with gonophores growing on a *Halocordyle* stem; *gon*, gonophore.

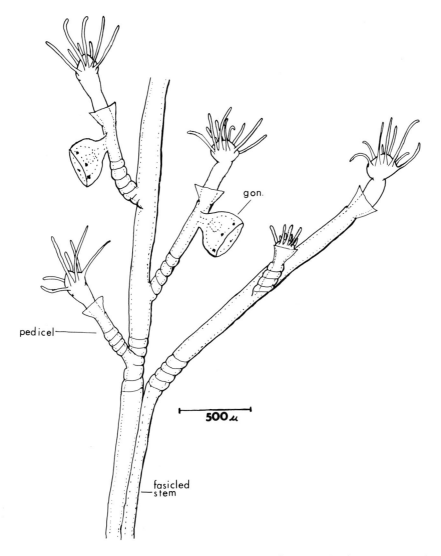

Figure 13.—HYDROIDA. *Bougainvillia ramosa,* a portion of a colony, showing polyps, gonophores containing medusae, and the fasicled lower stem; *gon,* gonophore.

mm in size, adding 2 to 3 more tentacles to each marginal cluster, while the oral tentacles branch once or twice. This is probably the hydroid called *Bougainvillia* sp. by Tusov and Davis (1971). Their material was collected from Kaneohe Bay, Oahu, as was the present material, which often grew as dense colonies on *Halocordyle* stems. Both the polyp and medusa stage were observed in this material. This is a generally distributed hydroid, more common in temperate waters, but known to penetrate subtropically to Brazil and Australia.

Family **Eudendriidae**

This family is characterized by a sharply differentiated hypostome, usually described as trumpet-shaped. There is a basal ring of filiform tentacles. Reproduction is always by means of an attached gonophore. *Eudendrium* is the only genus in the family.

Eudendrium sp. The hydranths of this hydroid usually arise singly from a basal stolon but may form colonies as high as 5 mm (Fig. 14). The polyps are light brown in color, about 0.5 mm high, with a ring of 15 or so tentacles. The perisarc bears a few annulations on some branches or is completely annulated on others. The annulations are very light and often indistinct. This hydroid is similar to *E. capillare,* but since no gonophores were observed, a specific identification is not possible. Specimens have been collected from Honolulu Harbor and Kaneohe Bay, Oahu.

Suborder CALYPTOBLASTEA

Family **Haleciidae**

The family Haleciidae is characterized by nonoperculate hydrothecae which may be simple off-shoots of the stem, or which may have a short inner wall, but which never have the base of the hydrotheca delineated by an internal wall. The polyps are often only partially retractile into the hydrothecae. The gonophores are fixed sporosacs, and the female gonotheca are often surmounted by two polyps. Some members of the family bear dactylozoids, but those in the genus *Halecium* do not.

Halecium beani (Johnston 1838). This species forms small colonies up to 20 mm high with sparse, irregular branching. The lower part of the stem is fasicled, but most of the stem is simple and slightly zigzag (Fig. 15). The stem internodes are from 0.5 mm to 0.8 mm long, and have a midinternode diameter of 0.2 mm. A hydrotheca is borne at the distal end of each internode. The hydrothecae are alter-

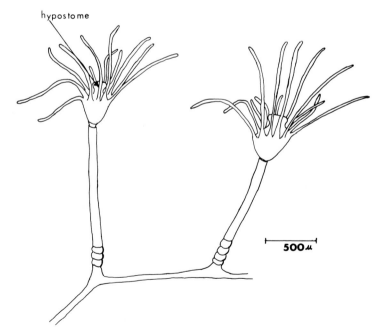

Figure 14.—HYDROIDA. *Eudendrium* sp., a view of two polyps from life, showing the characteristic hypostome.

nately directed, although the axis of the stem may be slightly twisted so that all hydrothecae do not necessarily lie in the same plane. The hydrothecae consist of a short hydrophore arising from the stem directly, and distally, a collar. This collar is usually marked with a circle of small perforations, the punctae. The diameter of the hydrophore is approximately 0.2 mm and the collar flares slightly so that the diameter of the aperture is 0.22 mm to 0.24 mm. All the material observed was sterile.

The species has been reported to be common enough locally to cause a number of cases of dermatitis (DeOreo, 1946, cited in Chu and Cutress, 1955), although the only material I have examined was from rubble at a depth of 30 feet in Maunalua Bay. While *H. beani* is a very cosmopolitan species, more common in temperate than in tropical waters, it has recently been found at Enewetak (Cooke, 1975).

Family **Campanulariidae**

This family is characterized by having a cup-shaped (campanulate) theca into which the polyp may be withdrawn. The polyp has a ring of filiform tentacles and a distinct hypostome. Reproduction may be by means of attached gonophores or by several sorts of medusae, depending on the genus. Many species in this family are insufficiently described, and more work is needed to determine the limits of valid species, making it somewhat difficult to assess the Hawaiian forms. Two forms with

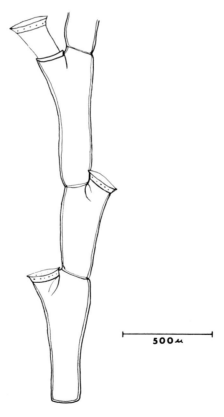

Figure 15.—HYDROIDA. *Halecium beani,* a portion of the stem perisarc showing the short hydrothecae, reduplicate in one case, and the collar punctae.

reasonably certain identifications are included here—*Obelia dichotoma* (Linnaeus) and *Clytia hemisphaerica* (Linnaeus)—and several other forms, including other *Clytia* and perhaps an *Orthopyxis,* have been seen.

Obelia dichotoma (Linnaeus 1758). The colonies consist of upright stalks, irregularly branched, if at all, as tall as 2 cm, arising from a creeping hydrorhiza. These stalks bear hydrothecal pedicels alternately. The pedicels may be covered with different degrees of annulation (Fig. 16). The hydrothecae are 0.3 mm high by 0.25 mm in diameter. The rims may appear entire or may seem to be made of small plates butted together. The gonothecae, shaped like high funnels with a raised lip, and a truncated, conical aperture arise from the main stalk at the axes. Medusae with 8 to 16 tentacles are often visible within the gonothecae. This is a cosmopolitan species, probably synonymous with *O. australis* von Ledenfeld from the South Pacific. Hawaiian specimens have been collected from Kaneohe Bay.

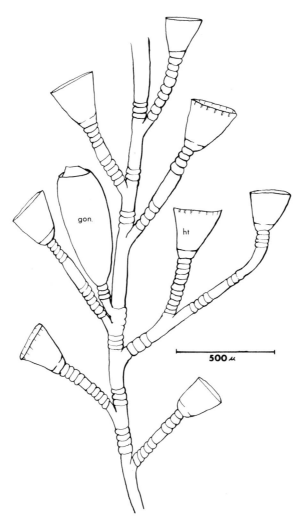

Figure 16.—HYDROIDA. *Obelia dichotoma,* a portion of the colony, showing hydrothecae and a gonotheca; *gon,* gonotheca; *ht,* hydrotheca.

Clytia hemisphaerica (Linnaeus 1767). Sterile colonies of what appear to be this species were collected from many different sites in Kaneohe Bay. The hydrothecae are 0.5 mm to 0.6 mm tall by 0.25 mm to 0.3 mm in diameter at the margin. The margin is marked by approximately 12 rather acute teeth separated by deep, round indentations (Fig. 17). These hydrothecae are borne on pedicels 1 mm to 3 mm long which usually have 10 to 15 annulations at the base and 3 or 4 just below the hydrothecae. Although absolute certainty is not possible with sterile material, this sample fits the descriptions of other authors extremely well. This is another cosmopolitan species which is known to show a great deal of variation in size and aspect under different ecological conditions.

Family **Campanulinidae** (?)

Hydroids of this family are characterized by having tubular hydrothecae which are closed by a conical operculum of many small flaps. These hydroids can give rise to medusae currently assigned to many different genera in several families. Moreover, there is considerable variation in the size and habit of the polyps from the same species. One member of this group has been observed in Hawaii.

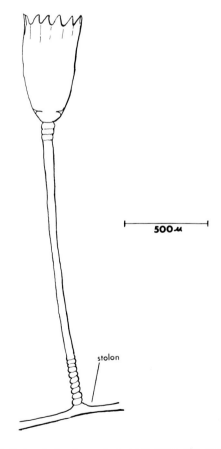

Figure 17.—HYDROIDA. *Clytia hemisphaerica*, a single hydrotheca with pedicel and portion of stolon.

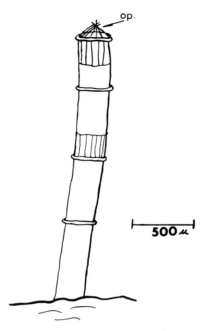

Figure 18.—HYDROIDA. *Cuspidella* sp., a single tubular hydrotheca arising from the substrate; *op*, operculum.

Cuspidella sp. Colonies of this hydroid consist of individual polyps arising singly from a subsurface hydrorhiza. The hydranths are enclosed in a transparent tubular hydrotheca, which is usually straight but may be slightly curved (Fig. 18). These hydrothecae are up to 2 mm long, and 0.2 mm in diameter, though they are often smaller (but with the same proportions). The hydrothecae are horizontally ridged several times, in the manner of *C. costata* (Hincks) and may have a section with light vertical lines immediately below the operculum. The operculum is composed of approximately 15 small triangular flaps which join inexactly to give a somewhat tufted appearance when closed. The operculum may be inverted into the theca if the theca is empty. The hydranth is almost colorless, has a conical hypostome, and a single basal whorl of filiform tentacles, approximately 14. When fully extended, the hydranth, supported on a long stalk, may reach another 1 mm beyond the end of the hydrotheca. Since these hydroids can give rise to medusae from several families, the exact position of any species is rather uncertain unless it has been studied through a complete life cycle. The material described was collected from the Kahe Point reef, Oahu. It was attached to coral rubble at a depth of 2 meters.

Family **Syntheciidae**

This family is characterized by nonoperculate hydrothecae which are usually more or less adnate to the stem. The hydrotheca is also marked by a basal diaphragm. Usually erect and branched colonies are formed. The gonophores are fixed sporosacs which often arise from the interior of a hydrotheca.

Synthecium tubitheca (Allman 1877). This species forms colonies of erect stems 1 cm to 2 cm high with several pairs (at most) of oppositely arranged side branches. Most stems are smaller and unbranched. These are indistinctly divided by faint nodes including one that defines a basal athecate portion. The hydrothecae are

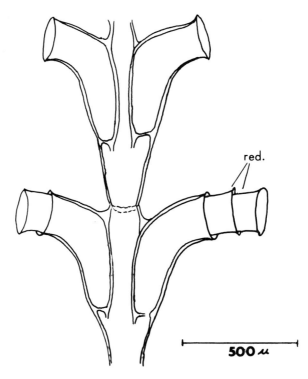

Figure 19.—HYDROIDA. *Synthecium tubitheca,* a portion of the stem showing two pairs of hydrothecae, one pair with reduplicated apertures.

oppositely arranged or just slightly misaligned (Fig. 19). They are tubular throughout, with never any contact between the walls of the pair. The length of the adnate portion is 0.3 mm to 0.5 mm and the length of the free portion between 0.2 mm to 0.3 mm. The margin is distinctly circular, with no teeth, but with reduplication in some cases. The marginal diameter is 0.15 mm to 0.20 mm. The distance between bases of consecutive hydrothecal pairs is 0.8 mm to 1 mm. No gonophores were observed on this material. This species (which may at first be taken for a small sertularid) is cosmopolitan, being widely reported from the tropical parts of the Atlantic, Pacific, and Indian Oceans. The local material was collected from 1 m to 2 m in Kaneohe Bay, where it was moderately common on coral rubble.

Family **Sertulariidae**

This family is characterized by having operculate thecae which are adnate to or immersed in the stem or branches. Fixed sporosacs are produced, and the gonothecae are often highly sculptured. There are at least six members of the family present in Hawaii, probably more, since they tend to be quite small and delicate, and are thus often overlooked. The following key will help separate those described below, but any questionable specimen should be checked against the descriptions and other literature.

KEY TO SERTULARIDS KNOWN FROM HAWAIIAN WATERS

1 Hydrothecae alternate or semiopposite 2
   Hydrothecae always strictly opposite 3

2(1)   Hydrothecae with complete adcauline wall and sharply curved free
       portion .... *Dynamena crisioides*
       Hydrothecae with incomplete adcauline wall, short slightly curved free
       portion .... *Sertularella speciosa*
3(1)   Hydrothecae smooth-walled 4
       Hydrothecae with many horizontal ridges .... *Sertularia subtilis*
4(3)   Interior of hydrothecae free of ridges or septa 5
       Hydrothecae with an internal ridge projecting from abcauline
       wall .... *Sertularia ligulata*
5(4)   Small unbranched colonies with very small (0.3 mm long), delicate
       hydrothecae, gonothecae smooth .... *Sertularia distans gracilis*
       Robust, branched colonies with hydrothecae approximately 0.7 mm
       long, gonothecae ridged .... *Dynamena cornicina*

Dynamena crisioides Lamouroux 1824. This hydroid forms colonies to 2 cm in height, with often just an unbranched main stem, or with alternating side branches. The hydrothecae are alternate to subopposite, and the branches have a thecate internode where they join the stem (Fig. 20). The hydrothecae are about 0.5 mm long, the top quarter being free and curving outward. The rim of the aperture has three teeth, although these may be obscure. The rim may also be reduplicated. There are three opercular flaps—one abcauline, and two adcauline—although these are often difficult to resolve (as are other details) because specimens are usually covered with

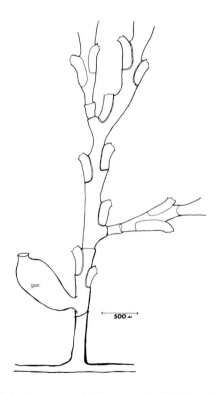

Figure 20.—HYDROIDA. *Dynamena crisioides,* a portion of the basal part of a colony and stolon, showing the manner of branching, position of the hydrothecae, and a gonotheca; *gon,* gonotheca.

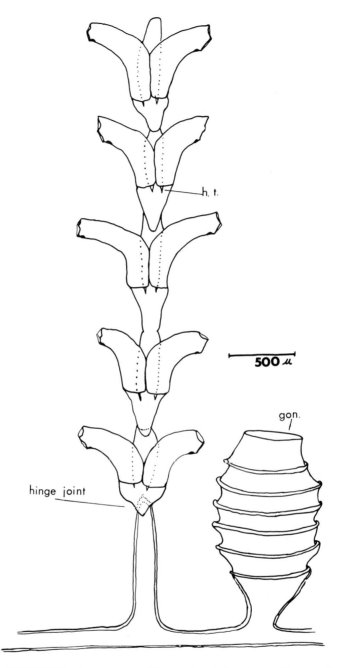

Figure 21.—HYDROIDA. *Dynamena cornicina*, a view of the colony with a gonotheca arising from the stolon; *ht*, hydrothecal teeth.

detritus and algae. Some of these conditions are variable (for example, size, degree of curvature of the hydrothecae) but the shape of the gonothecae is relatively constant. They are swollen oval structures about 0.75 mm to 1.0 mm long with a flared lip. Often some of these arise from hydrothecae. This species is found throughout the tropics in the Atlantic, Indian, and Pacific Oceans. Local specimens were collected from Kewalo Basin and Honolulu Harbor, Oahu, where they are common on substrates of rock and rusting iron.

Dynamena cornicina McCrady 1858. This hydroid usually forms colonies with many unbranched stalks, 1.0 mm to 1.5 mm high rising from a creeping stolon, though some stalks may branch. The stem is divided below the first pair of hydrothecae by an oblique hinge joint (Fig. 21). This basal athecate portion is of variable length. Above this portion the hydrothecae are borne in pairs on internodes which are separated by circular constrictions (which may be indistinct). The hydrothecae are more or less cylindrical with a diameter of 0.1 mm. They are adnate for approximately 0.5 mm and then curve outward for between 0.2 mm and 0.4 mm. Members of a pair are contiguous for some distance on the front of the stem but separated in back. The margin of the hydrothecal aperture bears three teeth—two lateral and one

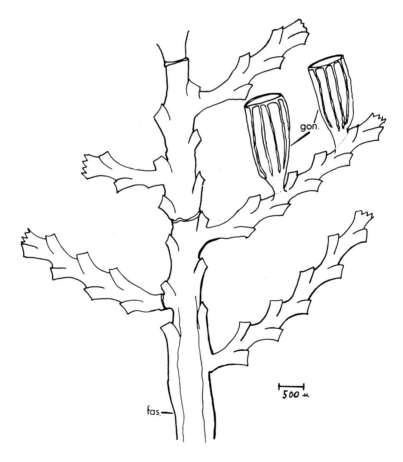

Figure 22.—HYDROIDA. *Sertularella speciosa*, a portion of a colony, showing the fasicled lower stem, and the gonophores arising from the branches; *fas,* fasicled stem; *gon,* gonothecae.

adcauline—and is slightly thickened under the abcauline rim. The operculum is composed of two flaps—a large abcauline and curved adcauline. The proximal walls of the hydrothecae bear teeth, usually two, which project into the interthecal space below. The gonothecae, even within a single colony, are rather variable in detail, but follow the same general design. They are ovoid or somewhat cylindrical, 1 mm to 2 mm long, and approximately 0.5 mm in broadest diameter. They are ringed by between six to nine ridges, which may be mere thickenings, or may be distinctly sculptured rings. The gonothecal aperture is circular and simple with no teeth. The gonothecae usually arise from the basal stolons in the vicinity of the stems, but some are borne on the lower parts of some stems. One example was seen of a gonotheca arising from a hydrotheca as in *Dynamena crisioides,* but this is evidently quite rare. This is a cosmopolitan species widely recorded from the Atlantic, Pacific, and Indian Oceans. The present material was collected from Coconut Island in Kaneohe Bay, Oahu.

Sertularella speciosa Congdon 1907. This species forms a colony with an erect, fasicled main stem and alternately arranged side branches. Colonies may reach 20 cm. The upper portion is nonfasicled and is divided into internodes which have two hydrothecae and a branch between them on one side and one hydrotheca opposite the branch joint (Fig. 22). The side branches are simple, each having a distinct node near the joint, and are divided into internodes which bear a variable number of hydrothecae. The hydrothecae are deeply immersed in the stem or branch and are

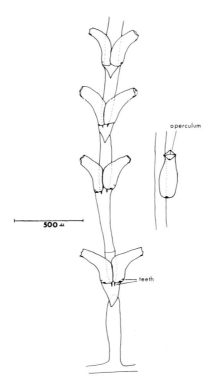

Figure 23.—HYDROIDA. *Sertularia distans gracilis,* a portion of the colony, with a lateral view of one hydrotheca showing the operculum.

strictly alternate, the aperture of the lower theca being just level with the base of the adcauline wall of the next higher alternate hydrotheca. The adcauline wall is incomplete and curved, while the abcauline wall is straight. The length of a hydrotheca is 0.4 mm, and the diameter is 0.2 mm to 2.3 mm. The margin has four small triangular teeth and a thickened ridge. The operculum (rarely intact) is four small plates. The gonothecae are borne on short stalks on the upper surfaces of the side branches, often two to four on a branch. They are cylindrical in outline, about 1.5 mm long and 0.5 mm to 0.6 mm in diameter. The surface is sculptured by thick vertical ribs, and there is a distinctly thickened rim at the top. The specimen examined agrees with the descriptions of *S. speciosa* quite completely, although that species is known only from the western Atlantic. *S. speciosa* is most probably synonymous with *S. diaphana*, from Australia. The local material was collected from shallow water in Kaneohe Bay, Oahu.

Sertularia distans (Lamarck 1816) *gracilis* Hassall 1848. This exceedingly small and delicate sertularid grows as unbranched colonies 4 mm tall. Upright stems, 0.05 mm to 0.06 mm in diameter, arise from a creeping stolon, with a basal athecate portion separated by a diagonal hinge joint from the first thecate-bearing internode. The internodes are separated by circular nodes and usually have one or two pairs of strictly opposite hydrothecae (Fig. 23). The hydrothecae are contiguous in front approximately 0.10 mm to 0.15 mm, while separated on the back of the stem. The attached length of the hydrothecae is 0.14 mm to 0.16 mm while the free portion is 0.16 mm to 0.19 mm long. The diameter of the hydrothecae is approximately 0.08 mm, perhaps a bit more at the base. There is approximately 0.3 mm to 0.4 mm between consecutive pairs of hydrothecae. The operculum consists of two flaps—one curved adcauline and one platelike abcauline. Proximally, the hydrotheca bears teeth which may project into the hydrotheca from the abcauline wall, upward from the lower wall, or downward into the interthecal space of the stem. This is a cosmopolitan species with this form *(gracilis)* present in many tropical locations. Local material was collected from coral rubble, 2 m deep, from the Kahe Point reef, Oahu.

Sertularia ligulata (Lamouroux 1816). This species forms colonies up to 10 mm tall, which are usually unbranched, though in some instances branches may arise from the basal portion of the stem. The stem is separated from the stolon by a short athecate internode, above which there are internodes each bearing one opposite pair of hydrothecae (Fig. 24). All nodes are oblique hinge joints. The diameter of the main stem is 0.08 mm. The hydrothecae are sharply curved with approximately 0.2 mm of adnate adcauline wall and 0.15 mm to 0.18 mm of free adcauline wall. The hydrothecae are contiguous for 0.1 mm to 0.2 mm of the front. The hydrothecae are distinguished by the presence of an internal peridermal ridge which projects into the hydrotheca from the abcauline wall. The hydrothecal aperture is approximately 0.1 mm in diameter and has two lateral teeth and a small adcauline tooth, though these exhibit considerable variation owing to age and wear. The operculum consists of one large abcauline flap and one small adcauline flap. Occasionally there are weak teeth projecting from the base of the hydrothecae into the interthecal space, although these are not always present. This species is widely distributed in the tropical parts of the Atlantic, Pacific, and Indian Oceans. The local material was collected from the surface of *Porites lobata* in 2 m in Kaneohe Bay, Oahu.

Sertularia subtilis Fraser 1937. This hydroid forms colonies consisting of upright, unbranched stems up to 10 mm tall. The hydrothecae are borne on the upper

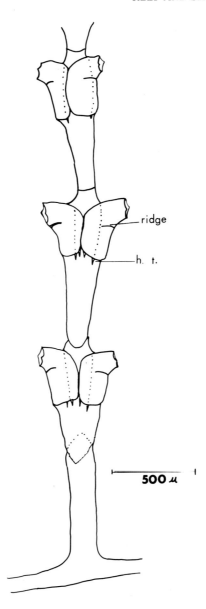

Figure 24.—HYDROIDA. *Sertularia ligulata*, a portion of a colony showing the hydrothecae with the internal ridge; *h.t.*, hydrothecal teeth.

two-thirds of the stem, the basal portion being an athecate length of stem separated from the stolon by a transverse annulation (Fig. 25), and from the first thecate internode by a diagonal hinge joint. The hydrothecae are strictly opposite and contiguous along their front face. They are approximately 0.25 mm long, with an adnate portion 0.5 mm long, and a free length of 0.18 mm. The diameter of the aperture is 0.1 mm. The distance between consecutive hydrothecae is from 0.30 mm to 0.36 mm. The hydrothecae are strongly and dramatically ringed by approximately 20 horizontal ridges. The operculum is composed of two flaps—a flat abcauline plate and a jointed adcauline one. Four marginal teeth were observed. The local material was compared

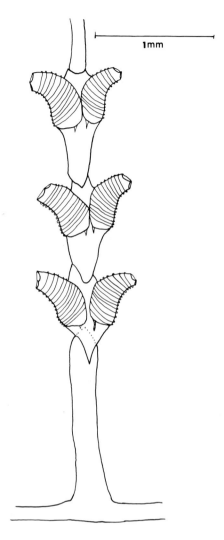

Figure 25.—HYDROIDA. *Sertularia subtilis,* a portion of a colony.

with the type specimen (USNM #43288) from Fraser's collection off Puerto Rico, its only previously reported location. The specimens were identical with the exception of the number of marginal teeth, and I thus considered our material to be of that species. The local material was collected from coral rubble 2 m deep on the Kahe Point reef, Oahu.

Family **Plumulariidae**

This family is characterized by the presence of the hydrothecae on the upper surfaces of the branches only and the presence of specialized, nonfeeding polyps—nematophores—encased in nematothecae. Reproduction is by means of fixed sporosacs, which are often protected by elaborate constructions. Three members of the family have been observed here, but there are most probably more.

Halopteris diaphana (Heller 1868). The colonies of this hydroid consist of stems about 1 cm to 1.5 cm long arising from the hydrorhiza (Fig. 26). The side

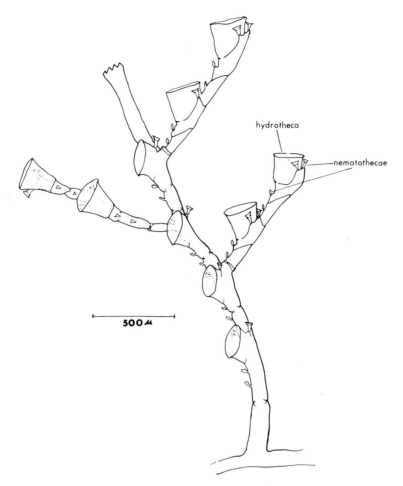

Figure 26.—HYDROIDA. *Halopteris diaphana*, a portion of a colony showing hydrotheca and nematothecae.

branches arise alternately, and a hydrotheca is present on the main stem on each internode which gives off a side branch. The branches are divided by septa into internodes about 0.5 mm long which bear the hydrothecae and nematothecae. The hydrothecae are about 0.3 mm long and 0.2 mm in diameter at the aperture. The abcauline wall is more or less straight and the rim of the aperture evenly circular. The nematothecae are arranged with two larger movable nematothecae on either side of the hydrotheca and one or two small ones behind the hydrotheca. The branches also bear internodes with just one nematotheca. No gonothecae were seen on the local material. This species is cosmopolitan in tropical seas. The local material was collected in Kaneohe Bay, Oahu, from a variety of substrates.

Plumularia margaretta (Nutting 1905). This species forms small delicate colonies of single stems 1 mm to 2 mm high arising from a loose network of creeping stolons. The main stem rises in a zigzag manner with a side-branch bearing one hydrotheca given off at each bend (Fig. 27). The single hydrotheca per branch is the distinguishing characteristic of this species. These side branches also include paired nodes which appear like truncated cones stuck one into the other. There are usually

HYDROZOA

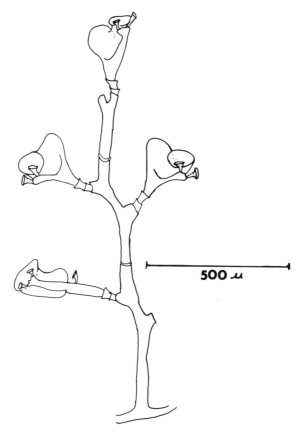

Figure 27.—HYDROIDA. *Plumularia margaretta*, a complete colony, with part of the stolon.

one or two nematothecae in the axis of the branch, one medially on it, often one over the front of the hydrotheca, as well as a Y-shaped pair arising from the end of the branch appressed to the back of the hydrotheca. The hydrothecae are usually recurved along the axis of the branch so that the margin, rather than the base, is closer to the main stem. The basal portion of the hydrotheca is inflated so that it sometimes appears as a separate chamber. The diameter at the aperture is approximately 0.1 mm. No gonothecae were found on the present material. This species is widely distributed in the tropical Atlantic, but the only Pacific records are from the west coast of Central America. The local material was collected growing on the red alga *Amansia* at Kahe Point reef, Oahu.

Plumularia setacea (Linnaeus 1758). This species forms colonies up to 20 mm high consisting of a simple main stem and alternate side branches (Fig. 28). The main stem is approximately 0.08 mm to 0.16 mm in diameter and is divided into internodes approximately 0.24 mm long. On each internode, the origin of a side branch is marked by a small projection bearing a single nematotheca. This immediately joins a small (0.06 mm to 0.08 mm) athecate internode of the side branch, which is further divided into hydrothecate and nonhydrothecate internodes. The internodes without a hydrotheca bear just one nematotheca. The hydrothecae are approximately 0.08 mm long and 0.08 mm to 0.09 mm in diameter with the aperture being adnate to the branch. There is one nematotheca behind the hydrotheca and a

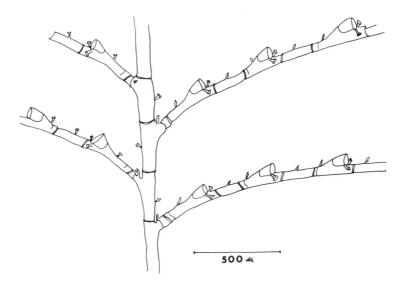

Figure 28.—HYDROIDA. *Plumularia setacea*, a portion of the colony about mid-stem height, showing the branching, hydrothecae, and distribution of nematothecae.

pair beside it. All these nematothecae are between 0.05 mm and 0.07 mm long and have a diameter of approximately 0.05 mm. No gonothecae were observed. This is truly a cosmopolitan species, being found in both temperate and tropical parts of all oceans. This species (or one similar) is likely to be that described by Edmondson (1946, Fig. 13d), since it is neither of the other two plumularids. The present material was collected from a variety of substrates in Kaneohe Bay, Oahu.

GLOSSARY (HYDROIDA)

abcauline: away from the stem.
adcauline: next to the stem.
adnate: with one side adherent to a stem or branch.
annulations: rings on stem or pedicel perisarc which allow flexibility.
aperture: opening of the perisarc through which a polyp emerges (may or may not be covered).
athecate: without a perisarcal covering for the retracted polyp.
bell: main body of a medusa.
calyptoblastea: suborder of thecate hydroids (Leptomedusae).
campanulate: bell-shaped.
capitate: with an enlarged terminal ball of nematocysts.
distal: farthest from the base.
fasicled: having several stems bound together.
filiform: decreasing gradually in diameter distally, with an even nematocyst distribution along the length.
gonophore: reproductive polyp.
gonotheca: perisarcal covering of a gonophore.
gymnoblastea: suborder of thecate hydroids (Anthomedusae).
hydranth: feeding polyp.
hydrophore: a shelf of perisarc at the base of the polyp.
hydrorhiza: attachment stolons of a colony which spread over the substrate.
hydrotheca: perisarcal covering of a hydranth.
hypostome: upper, highly extensible portion of the hydranth bearing the mouth.
manubrium: hollow, pendulous structure supporting the mouth of a medusa.
moniliform: with nematocysts localized in discrete rings along the tentacle length.
nematophore: mouthless defensive polyp with many nematocycts.
nematotheca: perisarcal covering of the nematophore.

node: joint in the stem or branch.
ocellus: eyespot of a medusa.
pedicel: stalk and perisarc supporting a polyp away from stem or branch.
perisarc: chitinous covering secreted by the ectoderm.
proximal: closest to the base.
sessile: attached directly to stem or branch.
sporosac: fixed reproductive structure.
stolon: basal tubular extensions of the hydroid stem attached to the substrate.
thecate: with a portion of the perisarc covering the retracted polyp.

REFERENCES (HYDROIDA)

Boone, L.
    1938. Coelenterata: Hydroida. Scientific Results of the World Cruises of the Yachts *Ara,* 1928–1929, and *Alva,* 1931–1932, *Alva* Mediterranean Cruise, 1933, and *Alva* South American Cruise, 1935, William K. Vanderbilt, Commanding. *Bull. Vanderbilt Marine Mus.* 7:33–34.

Chu, G. W. T. C., and C. E. Cutress
    1954. Human Dermatitis Caused by Marine Organisms in Hawaii. *Proc. Hawaiian Acad. Science* (1953) 54: 9.
    1955. Dermatitis Due to Contact with the Hydroid *Syncoryne mirabilis* (Agassiz, 1862). *Hawaii Medical J.* 14(5): 403–404.

Cooke, W. J.
    1975. Shallow Water Hydroids from Enewetak Atoll, Marshall Islands. *Micronesica* 11(1): 85–108. 6 pls.

De Oreo, G. A.
    1946. Dermatitis Venata Resulting from Contact with Marine Animals (Hydroids). Report of Cases. *Arch. Dermatology Syphilology* 54(6): 637–649.

Edmondson, C. H.
    1930. New Hawaiian Medusae. *B. P. Bishop Mus. Occ. Pap.* 9(6): 1–16.
    1946. *Reef and Shore Fauna of Hawaii.* B. P. Bishop Mus. Spec. Publ. 22. 381 pp. Honolulu.

Hartlaub, C.
    1901. Hydroiden aus dem Stillen Ocean. Ergebnisse einer Reise nach dem Pacific (Schausland 1896–97). *Zoologische Jahrbuecher Abteilung Systematik* 4: 349–379.

Kramp, P. L.
    1952. Reports on the Lund University Chile Expedition 1948–49. 2. Medusae Collected by the Lund University Expedition 1948–49. *Acta Univ. Lund,* N.F. Avd. 2. 47(7): 1–19.
    1961. Synopsis of the Medusae of the World. *J. Marine Biological Assoc. U. K.* 40: 1–469.

Mammen, T. A.
    1963. On a Collection of Hydroids from South India. I, Athecata. *J. Marine Biological Assoc. India* 5(1): 27–61.
    1965. On a Collection of Hydroids from South India. II, Thecata Except Plumulariidae. *J. Marine Biological Assoc. India* 7(1): 1–59.

Mayer, A. G.
    1910. *The Medusae of the World,* Vols. 1, 2. Carnegie Inst. Washington Publ. 109. 55 pls.

Nutting, C. C.
    1900–1904. *American Hydroids. Part 1, Plumularidae, Part 2, Sertularidae.* Spec. Bull. U.S. National Mus. 100.
    1905. Hydroids of the Hawaiian Islands Collected by the Steamer *Albatross* in 1902. *Bull. U.S. Fish Commn.* 23(3): 931–959.

Pennycuik, P. R.
    1954. Faunistic Records from Queensland. Part V, Marine and Brackish Water Hydroids. *Univ. Queensland Pap. Dept. Zoology* Vol. 1, no. 6.

Rees, W. J.
    1958. The Relationships of *Moerisia lysoni* Boulenger, and the Family Moerisiidae, with the Capitate Hydroids. *Proc. Zoological Soc. London* 130: 537–545.

Tusov, J., and L. V. Davis
- 1971. Influence of Environmental Factors on the Growth of Bougainvillia Sp. In H. M. Lenhoff and others (eds.), *Experimental Coelenterate Biology,* pp. 52–65. Honolulu: Univ. Hawaii Press.

Uchida, T., and Z. Nagao
- 1959. The Life-History of a Japanese Brackish-Water Hydroid, *Ostroumovia horii. J. Faculty Science Hokkaido Univ., Ser. 6 Zoology* 14: 265–281

Vannucci, M.
- 1951. Hydrozoa e Scyphozoa Existentes no Instituto Paulista de Oceanográfia. *Bol. Inst. Paulista Oceanográfia* 2(1): 69–100.
- 1954. Hydrozoa e Scyphozoa Existentes no Instituto Oceanografico II. *Bol. Inst. Oceanografico Sao Paulo* 5(1): 95–149.

Vervoort, W.
- 1962. A Redescription of *Solanderia gracilis* Duchassaing and Michelin 1846, and General Notes on the Family Solanderiidae (Coelenterata: Hydrozoa). *Bull. Marine Science Gulf Caribbean* 12: 508–542.
- 1966. Skeletal Structure in the Solanderiidae and Its Bearing on Hydroid Classification. In W. J. Rees (ed.), *The Cnidaria and Their Evolution,* pp. 372–396. London: Academic Press.
- 1968. Report on a Collection of Hydroida from the Caribbean Region, Including an Annotated Checklist of Caribbean Hydroids. *Zoologische Verhandelingen Leiden* No. 92: 3–124. 41 figs.

OTHER HYDROZOANS

L. G. ELDREDGE and DENNIS M. DEVANEY

University of Guam and B. P. Bishop Museum

Order SIPHONOPHORA

SIPHONOPHORES are colonial coelenterates composed of both medusae and polyps budded from a common stem. The polypoid portions or individuals are the feeding gastrozooids, tactile dactylozooids, and reproductive gonozooids. The medusoid individuals are responsible for locomotion (swimming bells or nectophores), and reproduction (gonophores). A float or pneumatophore, once considered to be medusoid in origin, is also present in many forms and is composed of a double-walled chamber provided at the bottom with a gas gland. Some species have a float which serves as a hydrostatic organ, keeping the animal at a particular depth, while other species lack the pneumatophore altogether. *Physalia* floats on the surface and has its pneumatophore elevated (Fig. 1*a*). The nematocysts are particularly toxic in siphonophores. Only one species has been reported from shallow Hawaiian waters, although smaller forms are common in the oceanic plankton, including small colorless species sometimes observed in near-shore waters.

Physalia physalis (Linnaeus 1758) [as *Physalia utriculus* Eschscholtz in Edmondson, 1946]. Commonly known as the Portuguese man-of-war, this wind-blown wanderer frequently reaches the Hawaiian Islands, especially the windward coasts. While lacking a nectophore, it floats at the surface, with its pneumatophore forming an oval having a turgid crest which allows the animal to maneuver in the wind. When stimulated, powerful nematocysts, mainly in the tentacles of the dactylozooids, discharge a noxious and severely painful toxin. The float, in iridescent and bluish purple color, may be as long as 5 cm while the extended tentacles, of the same color, may reach well over 10 m. Beach-washed individuals are frequently eaten by the mole crab, *Hippa pacifica* (Bonnet, 1946; Matthews, 1955), and, when available, constitute one of this crab's more important sources of food. The ghost crabs, *Ocypode laevis* and *O. ceratophthalmus,* also scavenge on it (Fellows, 1966). While Totton (1960) maintained that *Physalia physalis* was the only species to be recognized worldwide, Cleland and Southcott (1965), following earlier workers, considered *P. utriculus* La Martinière to be an Indo-Pacific species having a single main tentacle, and *P. physalis* an Atlantic species with several main tentacles.

Order CHONDROPHORA

The chondrophores are also colonial hydrozoans. They lack an elongated stem and have the shape of a round or oval disk consisting of a multichambered float, from which hangs a single gastrozooid, surrounded by gonozooids that give

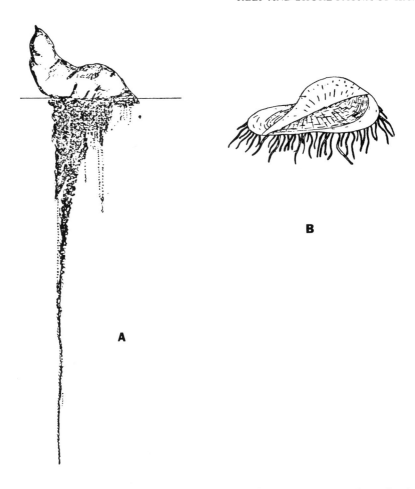

Figure 1.—OTHER HYDROZOANS: *a, Physalia physalis,* Portuguese man-of-war, floating on the water surface; note the single main tentacle (from Cleland and Southcott, 1965). *b, Velella velella* (from Hyman, 1940).

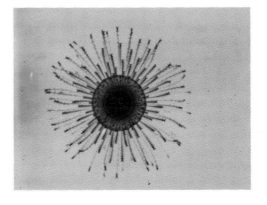

Figure 2.—OTHER HYDROZOANS: *Porpita pacifica* (C. E. Cutress photo).

rise to free-swimming medusae. Dactylozooids fringe the margin of the animal. Two members of this group of free-floating hydrozoans have been observed in Hawaiian waters. Neither has been reported to sting humans, in contrast to *Physalia*.

Velella velella Linneaus 1758 [as *V. pacifica* Eschscholtz in Edmondson, 1946]. This form has a flattened, elliptical float with a thin triangular sail (Fig. 1*b*). At the margin of the thin float is a series of short dactylozooids directed downward. Its nematocysts are less virulent than those of the man-of-war. The color of the animal is vivid blue. A review of this wide-ranging tropical species was given by Edwards (1966).

Porpita pacifica Lesson 1826. These small, discoidal forms, without a sail (Fig. 2), have been collected off the windward coast of Oahu. Specimens in Hawaiian waters seldom exceed a diameter of 1 cm to 2 cm. The marginally placed dactylozooids are quite evident, standing out horizontally at the side of the disk. Like *Velella,* this form is normally a blue color.

GLOSSARY (OTHER HYDROZOANS)

dactylozooid: tactile polyp which serves in defense and captures food.
gastrozooid: feeding polyp with mouth, tentacles, and gastric cavity.
gonozooid: reproductive polyp which buds medusae.
nectophore: swimming bell, a modified medusa with bell, velum, radial canals, and ring canal.
pneumatophore: float, an inverted medusoidlike bell.

REFERENCES (OTHER HYDROZOANS)

Bonnet, D. D.
    1946. The Portuguese Man-of-War as a Food Source for the Sand Crab *(Emerita pacifica)*. *Science* 103: 148–149.
Cleland, J. B., and R. V. Southcott
    1965. *Injuries to Man from Marine Invertebrates in the Australian Region.* Canberra: Commonwealth of Australia.
Edmondson, C. H.
    1946. *Reef and Shore Fauna of Hawaii.* B. P. Bishop Mus. Spec. Publ. 22. Honolulu.
Edwards, C.
    1966. *Velella velella* (L.): The Distribution of Its Dimorphic Forms in the Atlantic Ocean and the Mediterranean, with Comments on Its Nature and Affinities. In H. Barnes (ed.), *Some Contemporary Studies in Marine Science,* pp. 283–296. London: Allen and Unwin.
Fellows, D. P.
    1966. Zonation and Burrowing Behavior of the Ghost Crabs *Ocypode ceratophthalmus* (Pallas) and *Ocypode laevis* Dana in Hawaii. M.S. Thesis, Univ. Hawaii.
Hyman, L. H.
    1940. *The Invertebrates: Protozoa through Ctenophora.* New York: McGraw Hill.
Matthews, D. C.
    1955. Feeding Habits of the Sand Crab *Hippa pacifica* (Dana). *Pacific Science* 9: 382–386.
Totton, A. K.
    1960. Studies on *Physalia physalis* (L.). Part 1, Natural History and Morphology. *Discovery Rep.* 30: 301–367. Pls. vii–xxv, 31 figs.

CLASS SCYPHOZOA

DENNIS M. DEVANEY and L. G. ELDREDGE
B. P. Bishop Museum and University of Guam

THE SCYPHOZOANS include the more conspicuous types of jellyfishes, including most of the larger forms. While they are generally free-swimming medusae as adults (although one group of small forms retains a stalked condition), most pass through an attached larval stage, the scyphistoma, which, either through a process of transverse fission (strobilation) or total conversion of the polyp, produces free-living medusae. The margin of the bell (umbrella) may or may not be scalloped and can bear both nematocyst-containing tentacles and sensory bodies (rhopalia). With the exception of the order Cubomedusae and the semaeostomeid genus *Aurelia,* scyphozoan medusae lack the flap or ridge within the margin of the bell that characterizes many hydromedusae.

Four orders of scyphozoans have been reported from Hawaii in depths less than 25 fathoms. Open ocean forms are occasionally swept to near-shore waters by storms, with *Pelagia notiluca* (order Semaeostomeae) having washed up at Waikiki beach following several days of southerly Kona winds (C. E. Cutress, pers. comm.). Many jellyfishes can cause severe systemic reactions to bathers by their toxic stinging cells (nematocysts), and contact or handling should be avoided if possible. A comprehensive listing and taxonomic resumé of the Scyphozoa is given by Kramp (1961).

KEY TO SCYPHOZOANS KNOWN FROM HAWAII

1 Free-living, medusae (as adults) 2
   Permanently stalked, attached .... Order STAUROMEDUSAE
                                                    Kishinouyea hawaiiensis
2(1) Bell cuboid; with pedalia; margin not scalloped but infolded to form a
        marginal flap (velarium)................ Order CUBOMEDUSAE 3
     Bell not cuboid; without pedalia; margin usually scalloped into lappets
        and without velarium 4
3(2) Length of bell nearly twice that of width (maximum size, l:w = 80:50
        mm)... *Charybdea alata*
     Length of bell about same as width (maximum size, l:w = 35:30 mm)
        .... *Charybdea rastoni*
4(2) Mouth open and drawn out into four frilly oral lobes; marginal tentacles
        present (in genus *Aurelia*) .... Order SEMAEOSTOMEAE
                                                       Aurelia (?) *labiata*

	Mouth closed and oral lobes divided or complexly fused into eight or more mouth-arms (in adults) having numerous small grooves and canals in mouth-arms; marginal tentacles absent ....Order RHIZOSTOMEAE 5
5(4)	Benthic forms, bell often inverted and pulsating on bottom in shallow water; usually with short flattened or clublike appendages extending from mouth arms .... *Cassiopea* 6
	Free-floating, seldom inverted, often found near surface in quiet waters; elongate mouth-arms or naked appendages (tentacular filaments) extending from the mouth-arms 7
6(5)	Bell with aboral concavity; mouth-arm appendages leaflike. .... *Cassiopea medusa*
	Bell without aboral concavity; mouth-arm appendages clublike ....*Cassiopea mertensi*
7(5)	Central region of exumbrella bearing dome with conical protuberances .... *Cephea cephea*
	Exumbrellar surface of bell without protuberances 8
8(7)	Mouth-arms short terminating in long untapered appendages with a distal expansion;[1] with many light colored spots on bell............. .... *Phyllorhiza punctata*
	Mouth-arms long, tapered, distal filaments without appendages; bell surface without spots but margins with white areas .... *Thysanostoma flagellatum*

Order STAUROMEDUSAE

As attached medusae, members of this order are connected from the exumbrellar surface by a stalk held to the substrate by an adhesive disk.

Kishinouyea hawaiiensis Edmondson 1930 [as *K. pacifica* in Edmondson, 1946]. The stalk of this form is four-chambered. The bell margin consists of eight lobes fused in pairs, each terminating with 16 to 21 short capitate tentacles (Fig. 1). While alive, specimens are greenish brown. The height of the bell is about 6.5 mm. Specimens have been collected around Oahu on various species of algae in shallow water (Edmondson, 1930).

Order CUBOMEDUSAE

As the name suggests, species in this order have a cuboidal shape with long, thin tentacles borne on pedalia (thickened bases of the tentacles hanging down from the bell margin at the interradial corners of the bell). These organisms are characterized by having a velarium, which is an infolding of the lower rim of the bell, containing canals of the gastrovascular system. This group includes the sea wasp *(Chironex fleckeri),* a lethal species known in the Austro-Malayan region. A detailed discussion of the danger of the Cubomedusae including an account and distribution of the two Hawaiian representatives can be found in Cleland and Southcott (1965).

---

[1] The distal expansions often broken off leaving blunt tips.

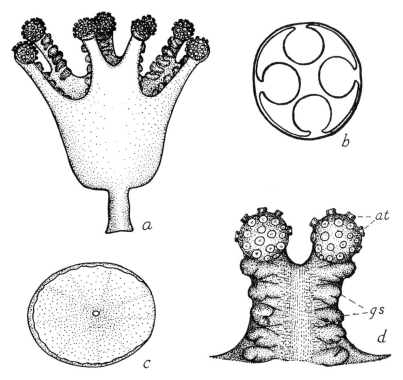

Figure 1.—SCYPHOZOA. *Kishinouyea hawaiiensis: a,* lateral view of the medusa; *b,* cross section of peduncle showing four chambers; *c,* basal end of peduncle showing central pore opening of a canal of the peduncle; *d,* medial surface of a lobe showing gonads and adhesive tentacles; *at,* adhesive tentacles; *gs,* gonads (from Edmondson, 1930).

Charybdea alata Reynaud 1830 [as *C. moseri* Mayer 1906]. In Hawaiian waters this species has been recorded up to 80 mm in height by 47 mm in width. Gonads are not developed until the bell reaches 30 mm to 60 mm in height. Four radially situated, club shaped sense organs, each with a wide cleftlike niche, occur about 15 mm above the margin of the velarium in large specimens (Fig. 2a). This species, like the next, is nearly transparent in water and difficult to see, but preserved specimens have shown the flexible shafts of the tentacles to be slightly pink, the eye spots dark reddish brown, and the gonads milky yellow (Mayer, 1906). Nearly all of the collected specimens have been found near the surface at various locations around the Hawaiian Islands. According to Edmondson (1952, p. 5), swarms of this jellyfish rather suddenly appeared on Waikiki Beach during June, 1951. The severity of the sting of this jellyfish is said to equal or exceed that of the Portuguese man-of-war *(Physalia). C. alata* is known from the tropical parts of the Atlantic, Pacific, and Indian Oceans. Life history studies at Puerto Rico were presented recently (Arneson and Cutress, 1977).

Charybdea rastoni Haacke 1886 [as *C. arborifera* Maas 1897]. In contrast to *C. alata,* the bell of *C. rastoni* is nearly as wide as high (maximum 30 mm by 35 mm). Gonads appear when the medusa bell is only 11 mm high and attain their maximum size when the bell height reaches 15 mm. The four club-shaped sense organs are found within a niche about 5 mm above the level of the velarium (Fig. 2b). Maas

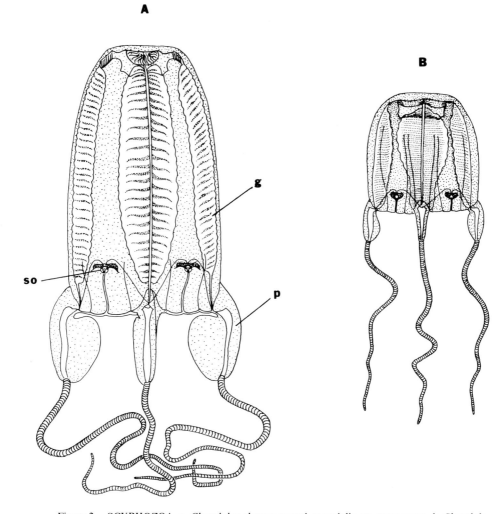

Figure 2.—SCYPHOZOA. a, *Charybdea alata;* g, gonad, p, pedalia, so, sense organ. b, *Charybdea rastoni* (both redrawn after Mayer, 1906, by David Kemble, Bishop Museum).

(1897) reported it from the surface waters at Honolulu. Mayer (1906) reported *C. rastoni* from depths of 23 fathoms or less around the islands of Maui and Kauai, and from surface waters near a Honolulu wharf using a night light. *C. rastoni* has been reported in the tropical Pacific, Japan, and S. Australia.

Order SEMAEOSTOMEAE

In this order, one unidentifiable species of *Aurelia,* possibly *A. labiata* Chamisso and Eysenhardt 1820, is seen, sometimes frequently, in harbors or inshore areas. It has 16 marginal lobes, and its radial canals branch into many parts (Fig. 3). The bell diameter attains 30 cm. When alive, the color is light—off-white to yellow—sometimes with a bluish tint. The gonads are crescent-shaped and opaque white to violet.

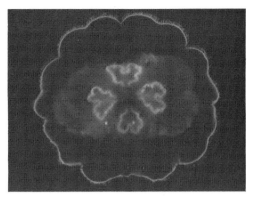

Figure 3.—SCYPHOZOA. *Aurelia* cf. *labiata* (photo by C. E. Cutress).

Order RHIZOSTOMEAE

All members of this order are characterized by the lack of marginal tentacles and a central mouth. During development the four corners of the mouth grow out and bifurcate several times to become mouth-arms. Continuous channels along the edges of these dendritic mouth-arms convey food to the stomach. The bell margin is scalloped into lappets. Five species of rhizostomes in four families are recognized from Hawaiian waters.

Family **Cassiopeidae**

Members of the genus *Cassiopea*, with two species reported from Hawaii, rest upside down on their bells in shallow water. By pulsating, they establish water currents that carry food to the mouth-arms. The length of the highly branched mouth-arms is usually not quite equal to the bell diameter. When raised off the bottom and floated right side up, individuals will turn over and pulsate to the bottom. Free swimming in this group is seldom observed but may occur during reproductive activity.

Figure 4.—SCYPHOZOA. *Cassiopea medusa*, aboral side (photo by C. E. Cutress).

Figure 5.—SCYPHOZOA. *Cassiopea medusa*, oral side (photo by C. E. Cutress).

Cassiopea medusa Light 1914. This species is known to reach 26 cm in bell diameter with mouth-arms 17 cm in length. The bell has an obvious aboral concavity in the central region, which is used as a sucker. There are between five and seven irregular and indistinct velar lappets between successive sense organs (rhopalia). The subdivisions of the mouth-arms have appendages varying in shape from linear to spatulate depending upon their position (Figs. 4, 5). It was reported by Chu and Cutress (1954) that only slight disturbance of the water around this species results in the detachment of many minute mouth parts containing nematocysts which remain viable in the water for days. The detached parts produced a dermatitis in rabbits and humans.

This species of *Cassiopea*, known originally from the Philippines, was considered as having been accidentally introduced to Hawaii by ships, first at Pearl Harbor between 1941 and 1945, and then spreading to Honolulu Harbor and the Ala Wai Canal by 1950 (Cutress, in Doty, 1961).

Cassiopea mertensi Brandt 1835. The bell of this species, reaching a maximum diameter of 20 cm, differs from *C. medusa* in lacking an aboral concavity. Eight tongue-shaped, prominently projecting lappets occur between the rhopalia. Eight cylindrical mouth-arms, less than the diameter but greater than the radius of the bell in length, give off 8 to 12 main branches and each of these divides dendritically. A number of club-shaped appendages occur between the mouth-arms.

Uchida (1970) first reported *C. mertensi* from the Hawaiian Islands on the sand in Kaneohe Bay at a depth of about 1 m. Extra-Hawaiian localities for *C. mertensi* are New Britain and the Caroline Islands.

Family **Cepheidae**

Cephea cephea (Forskål 1775). This free-swimming form attains a size of 100 mm to 140 mm in bell diameter. It is easily recognized, with the apex of the exumbrellar surface having a central dome covered by up to 30 large conical projections of unknown function (Fig. 6). Many frilled mouths are found on the lower sides of the mouth-arms and their branches. Up to a hundred or more long, tapering, pointed tentacular filaments come off the mouth-arms. The larger filaments (up to 16 in number) arise from the arm-disk at the points of origin of the eight mouth-arms, the smaller ones arising between the mouth-frills on the arms. Ocelli (pigmented eye spots) are absent from the sense organs. While the velar lappets are generally quite well developed, they are united by a web, giving the bell margin an appearance of being nearly entire. The ocular lappets are small and deeply set inward from the margin.

In April, 1975, a specimen of *Cephea* was observed at Waialea Bay on the northwest side of the island of Hawaii. A color photograph of this specimen appeared in an article by Mack (1975, p. 502) as *Cassiopea* sp. Specific determination as *C. cephea* was made by C. E. Cutress from photographs of the specimen (which was not collected). Another specimen was observed (Fig. 6) off the Kona coast of Hawaii. These are the first records of the species from Hawaiian waters.[2] This species is known from the Red Sea to southern Japan and as far east as the Gambier Islands in French Polynesia (Kramp, 1961).

---

[2]The citation in Mayer (1910, p. 655) which includes *Diplopilus couthouyi* (as a synonym of *C. cephea*) from Hawaii was inaccurate; *D. couthouyi* was reported from Wilson's Island, now known as Manihi in the Tuamoto archipelago.

Figure 6.—SCYPHOZOA. *Cephea cephea,* off Kona coast, island of Hawaii (photo by Chris Newbert).

Family **Mastigiidae**

Phyllorhiza punctata von Lendenfeld 1884 [as *Cotylorhizoides pacificus* (Mayer) by Cutress (in Doty, 1961); *Phyllorhiza pacifica* (Light) by Cutress (in Mansueti, 1963); *Mastigias ocellatus* (Modeer) in Walsh, 1967]. Species of *Phyllorhiza* have the mouth-arms triangular in cross section, the mouths occurring along the three edges as well as on the flat sides of the mouth-arms. In *P. punctata,* there extends from each mouth-arm an appendage of uniform thickness having an enlargement at the tip (often broken off, however). While the over-all color varies from Prussian blue to ocher, the bell is characteristically marked with rounded white spots (Fig. 7). The bell diameter reaches 50 cm, the mouth-arms are usually two-thirds as long as the bell diameter, and their appendages as much as two-thirds the length of the mouth-arms. This species is known to feed on microzooplankton.

This jellyfish has been reported harboring young carangid fish *(Caranx mate)* beneath the bell.[3] Juveniles of another carangid, *Gnathanodon speciosus,* have also been found with this jellyfish in Hawaii (Jeff Leis, pers. comm.).

Phyllorhiza punctata occurs, sometimes commonly, in Pearl Harbor, Honolulu Harbor, and Kaneohe Bay. The undetermined rhizostomid reported and figured by Edmondson (1933, 1946) as being abundant in Pearl Harbor at certain times of the

---

[3] The record of young cardinalfish, *Amia* (syn. *Apogon*) *frenatus,* as commensals from the body cavity of this rhizostomeid jellyfish in Pearl Harbor (Edmondson, 1946, pp. 29, 335) is considered a misidentification; they are not apogonids but "almost certainly *Caranx mate*" (Cutress, in Mansueti, 1963, p. 57).

year—mainly during the winter months—certainly appears to be *P. punctata.* Besides Hawaii, this species is reported from the tropical coastal waters of the Indo-Pacific and West Indian region (Cutress, 1973).

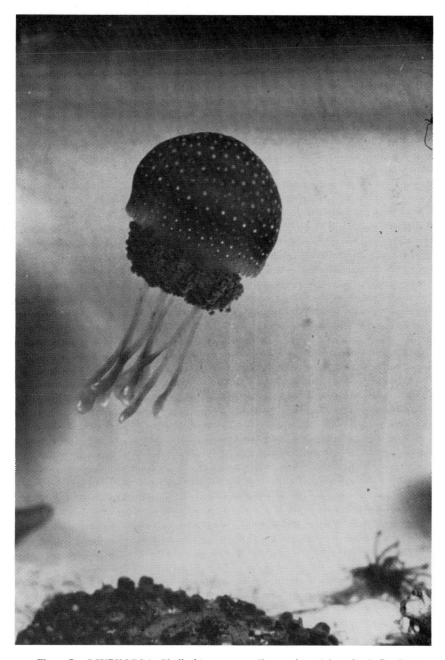

Figure 7.—SCYPHOZOA. *Phyllorhiza punctata,* live specimen (photo by A. Reed).

Family **Thysanostomatidae**

Thysanostoma flagellatum (Haeckel 1880). This jellyfish is characterized by having a somewhat flatly rounded bell, up to 20 cm wide and 5 cm high, and six to eight broadly rounded velar lappets between rhopalia (Kramp, 1961). The mouth-arms are 1 to 1.5 times the bell diameter and, while each of these has been considered to possess a narrow, filiform distal appendage (Stiasny, 1929), this was not mentioned in the original description, nor has it been observed on recent Hawaiian specimens.

Although originally described from specimens collected in the Hawaiian Islands in the 19th century, two smaller individuals of *T. flagellatum* were reported subsequently, one from the Philippines (Mayer, 1910) and one from the East Indies (Stiasny, 1929). In October, 1974, a rhizostomeid jellyfish was observed and photographed (but not collected) off Heeia pier in Kaneohe Bay near the surface (Fig. 8). The outer side of the bell appears granulated and has large white, irregularly shaped ovoid to rectangular pigmented areas located along the margins (Fig. 9). Small diffuse patches of brown algal cells are also evident near the margin especially over the marginal lappets where they contrast with the white pigment beneath. The mouth-arm appendages are broad with distinct whitish pigmented areas; no distal tapering filaments are evident. It seems likely that the specimen represents *T. flagellatum*, but future collections will be necessary to confirm this. Small carangid fish were observed beneath the bell.

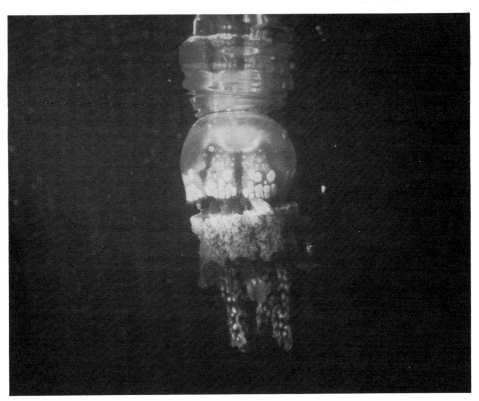

Figure 8.—SCYPHOZOA. (?) *Thysanostoma flagellatum*, off Heeia pier, Kaneohe Bay (photo by J. Grovhoug).

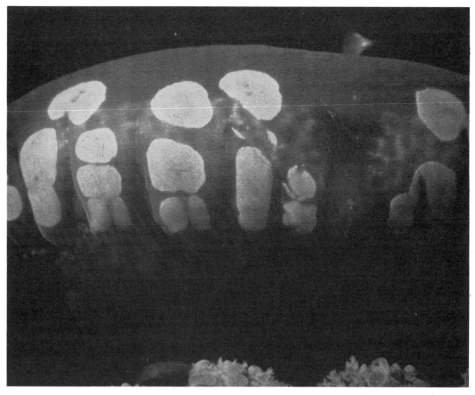

Figure 9.—SCYPHOZOA. (?) *Thysanostoma flagellatum*, as in Figure 8, showing margin of bell (photo by J. Grovhoug).

GLOSSARY (SCYPHOZOA)

aboral concavity: a hollow depression in the central exumbrellar surface of the bell used as a means of attachment to the substratum.
bell: see **umbrella.**
benthic: bottom dwelling.
capitate: knoblike; characteristic of the ends of the nematocyst-filled tentacles in the Stauromedusae.
dendritic: treelike branching.
dermatitis: an inflammation (rash) of the skin as caused by the nematocysts of jellyfish.
exumbrella: the top or aboral convex outer surface of the bell.
lappets: scalloped edges around the bell of some jellyfish **(marginal lappets): velar lappets,** those not bearing sense organs; **ocular lappets,** those bordering or containing the sense organs.
marginal lappets: see **lappets.**
mouth-arms: or oral arms, extensions of the oral lobes, often as frilly projections with or without appendages.
ocular lappets: see **lappets.**
oral lobes: four prolongations of the mouth.
pedalia: gelatinous expansions of the bell bearing one or more marginal tentacles and/or sense organs; especially in Cubomedusae.
pelagic: living in the open ocean not close to land.
radial canals: channels in the gastrovascular (digestive) system passing from the centrally placed stomach to the periphery of the bell.
rhopalia: marginal sense organs of the Schyphozoa containing a cystlike hydrostatic (balancing) organ (the statocyst) and sometimes light-sensitive cells (ocelli).

scyphistoma: the polypoid larva of Schyphozoa having four-part symmetry which becomes an adult directly or produces medusae (ephyra) by asexual reproduction.
sense organs: see **rhopalia.**
strobilation: asexual reproduction by transverse fission of the scyphistoma larva.
subumbrella: lower or inner side of the bell usually the concave side bearing the mouth.
umbrella: or **bell;** the discoidal, cuboidal, or hemispherical mass of a jellyfish; having ex- and subumbrellar surfaces.
velar lappets: see **lappets.**
velarium: a subumbrellar ridge found in Cubomedusae and *Aurelia.*

REFERENCES (SCYPHOZOA)

Arneson, C. A., and C. E. Cutress
    1977. Life History of *Carybdea alata* Reynaud, 1830 (Cubomedusae). In G. O. Mackie (ed.), *Coelenterate Ecology and Behavior (1976),* pp. 227–236. New York: Plenum.

Chu, G. W. T. C., and C. E. Cutress
    1954. Human Dermatitis Caused by Marine Organisms in Hawaii. *Proc. Hawaiian Acad. Science* 1953–1954, p. 9.

Cleland, J. B., and R. V. Southcott
    1965. *Injuries to Man from Marine Invertebrates in the Australian Region.* Canberra: Commonwealth of Australia.

Cutress, C. E.
    1973. *Phyllorhiza punctata* in the Tropical Atlantic. *Proc. Assoc. Island Marine Lab. Caribbean* 9: 14.

Doty, M. S.
    1961. *Acanthophora:* A Possible Invader of the Marine Flora of Hawaii. *Pacific Science* 15(4): 547–552. (Footnote by C. E. Cutress, p. 549.)

Edmondson, C. H.
    1930. New Hawaiian Medusae. *B. P. Bishop Mus. Occ. Pap.* 9(6): 1–16.
    1933, 1946. *Reef and Shore Fauna of Hawaii.* B. P. Bishop Mus. Spec. Publ. 22. Honolulu.
    1952. *Report of the Director for 1951.* B.P. Bishop Mus. Bull. 208. Honolulu.

Haeckel, E.
    1880. *System der Acraspeden. Zweite Hälfte des System der Medusen,* pp. 361–672. 20 pls. Jena.

Kramp, P. L.
    1961. Synopsis of Medusae of the World. *J. Marine Biological Assoc. U.K.* 40: 1–469.
    1970. Zoogeographical Studies on Rhizostomeae (Scyphozoa). *Videnskabelige Meddelelser Dansk Naturhistorisk Forening* 133: 7–30.

Maas, O.
    1897. Die Medusen. Report 23 on the Dredging Operations of the U.S. Steamer *Albatross* during 1891. *Mem. Mus. Comparative Zoology Harvard* 23: 1–92. 15 pls.

Mack, J.
    1975. Hawaii's First Natural Area Reserve. *Defender's Mag.* 50(6): 500–503.

Mansueti, R.
    1963. Symbiotic Behavior between Small Fishes and Jellyfishes. *Copeia* 1963: 40–80.

Mayer, A. G.
    1906. Medusae of the Hawaiian Islands Collected by the Steamer *Albatross* in 1902. *Bull. U.S. Fish Commn.* 23(3): 1131–1143.
    1910. *Medusae of the World.* Vol 3: *The Scyphomedusae.* Carnegie Inst. Washington Publ. 109, pp. 499–735, pls. 56–76. Washington, D.C.

Stiasny, G.
    1929. Ueber Einige Scyphomedusen aus dem Zoologischen Museum in Amsterdam. *Zoologische Mededelingen* 12: 195–215. 15 figs.

Uchida, T.
    1970. Occurrence of a Rhizostome Medusa, *Cassiopea mertensi* Brandt from the Hawaiian Islands. *Annotationes Zoologicae Japonenses* 43(2): 102–104. 2 figs.

Walsh, G. E.
    1967. An Ecological Study of a Hawaiian Mangrove Swamp. In George H. Lauft (ed.), *Estuaries,* pp. 420–431. American Assoc. Advancement Science Publ. 83. Washington, D.C.

CLASS ANTHOZOA

THE ANTHOZOANS are exclusively marine cnidarians which exist only as polyps. The oral area is expanded into a disk, and the body is, for the most part, relatively short. The body cavity is divided by septa into a number of chambers. The animal is either solitary or colonial and may or may not secrete a hard internal or external skeleton. The class is divided into two subclasses—the Octocorallia (or Alcyonaria) and the Zoantharia (or Hexacorallia)—primarily on the basis of symmetry.

Subclass OCTOCORALLIA

DENNIS M. DEVANEY
B. P. Bishop Museum

Polyps of this subclass possess eight pinnate (branched) tentacles and eight septa (membranes) which divide the body cavity into compartments. When present, the skeleton consists of individual sclerites or of a network of these fused together with either a calcareous or hornlike substance. Of the six recognized orders, five are found in Hawaiian waters, three in depths less than 100 m. No blue corals (order Coenothecalia) are present, while the orders Stolonifera and Pennatulacea (sea pens) are known only at depths exceeding 100 fm. Most of the other representatives of this subclass are deep-water forms (Bayer, 1952, 1956).

Order ALCYONACEA

Six families of this order, collectively referred to as "soft corals" are known, but only two are represented in Hawaiian waters. The predominant reef-dwelling alcyonaceans in the tropical Pacific are members of the family Alcyoniidae. However, until recently, this family was unreported in shallow Hawaiian waters. Alcyonaceans generally form massive or lobate colonies, occasionally arborescent, in which there is no solid axis, and in which the gastric cavities of many polyps reach to the base of the colony.

Family **Alcyoniidae**

Sinularia abrupta Tixier-Durivault 1970. Colonies of this true "soft coral" have been found at several locations in recent years around the island of Oahu. It has been observed off Moku Manu islet at a depth of 17 m to 37 m and off Sandy Beach and Rabbit Island (Manana) at depths less than 10 m. Verseveldt (1977) made the initial identification of these specimens and gave a redescription of this species. The colonies encrust hard substrata, either basalt or limestone. A single colony may extend nearly 1 m in greatest breadth. The general surface of the colony is lobose

Figure 1.—OCTOCORALLIA. *Sinularia abrupta* colony at Halona Blowhole, Oahu (courtesy of Michael Missakian).

with lobes from 1 cm to 5 cm high (Fig. 1), somewhat resembling poritid stony corals but being flexible to the touch. Each lobe consists of many polyps approximately 1 mm in diameter. Retracted, the polyps give the lobes a pitted appearance. The living color ranges from greenish to yellowish brown. Very well-developed rough elongate tuberculate sclerites are present as well as fusiform spindles (Figs. 2, 3). Besides Hawaii, *S. abrupta* is recorded from Nha-Trang, Vietnam, and Fanning Atoll.

Family **Xeniidae**

Anthelia edmondsoni (Verrill 1928) [as *Sarcothelia edmondsoni* Verrill]. This "soft coral" is occasionally found in shallow waters. The soft-bodied polyps, while close together, are only connected basally by a stolon and are 5 mm to 7 mm high when expanded. No sclerites are present. The tentacles are pale lilac to purple in reflected light, and the polyp body is light tan or buff (Figs. 4, 5). Recent studies have demonstrated the presence of zooxanthellae in the tissues of *A. edmondsoni* (Robert Kinzie, pers. comm.). Colonies reach a diameter of 8 cm or more and have been collected from various locations around the islands, generally in quiet waters or embayments. Utinomi (1950, 1958) brought attention to the fact that Verrill's *Sarcothelia* was synonymous with *Anthelia*. *A. edmondsoni* is known only from Hawaii.

Order TELESTACEA

This group is recognized by having erect, branching colonies which, in the family Telestidae, are produced by monopodial growth and in which each tall axial

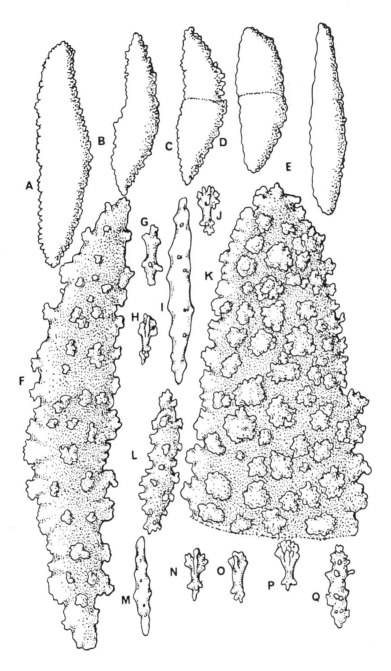

Figure 2.—OCTOCORALLIA. *Sinularia abrupta*. Fusiform (*a–e*) and tuberculate (*f–q*) spicules from capitular region (after Tixier-Durivault, 1970).

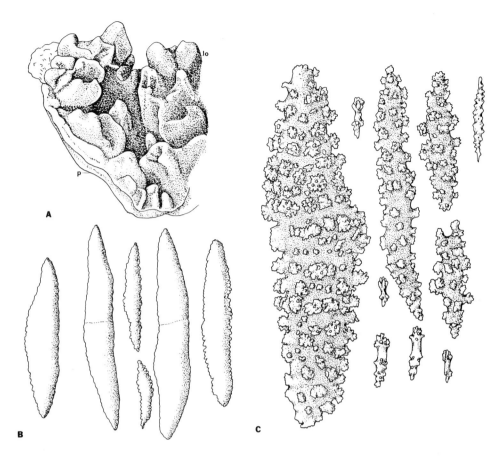

Figure 3.—OCTOCORALLIA. *Sinularia abrupta.* *a,* Portion of a colony (*lo,* lobe; *p,* base); *b,* fusiform sclerites from base; *c,* tuberculate sclerites from base (after Tixier-Durivault, 1970).

polyp has many short lateral polyps. In the genus *Telesto,* the polyps lack sclerites in the tissues of the lower part of the gastric cavities.

Telesto riisei (Duchassaing and Michelotti 1860). In the anthocodial region of the polyps of *T. riisei,* sclerites are longitudinally arranged in 16 narrow rows—8 on the tentacle basis and 8 along septal insertions below the tentacles (Fig. 6a). Two types of sclerites occur in the body wall (Figs. 6b-g). Colonies of this species (Fig. 7) are commonly red or orange and have white tentacles. In Hawaiian waters they were first noted in Pearl Harbor, Oahu (Naval Undersea Center, 1974). It occurs as a part of the fouling community, but was not noted by Edmondson in the 1940's during his studies in this location. Subsequent investigation has shown it to be present in Honolulu Harbor and several other areas along the leeward Oahu coast. The species has previously been reported from Florida to Brazil in the West Atlantic (Bayer, 1961) and it has quite likely been introduced to Hawaii.

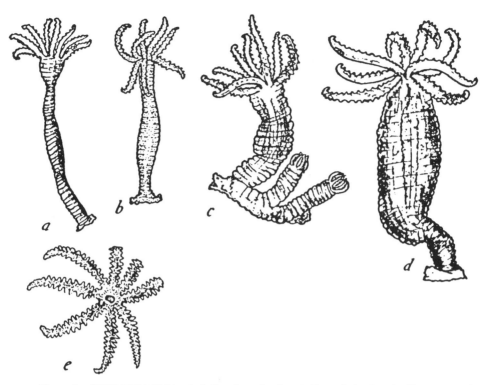

Figure 4.—OCTOCORALLIA. *Anthelia edmondsoni*. *a, b,* Expanded polyps (× 10); *c,* group of partially contracted polyps (× 15); *d,* partially contracted polyp (× 25); *e,* oral view of polyp (× 10) (from Verrill, 1928).

Figure 5.—OCTOCORALLIA. *Anthelia edmondsoni*. *a* (left), Colony from Sharks Cove, Oahu; *b* (right), same, close-up of polyps (courtesy S. A. Reed).

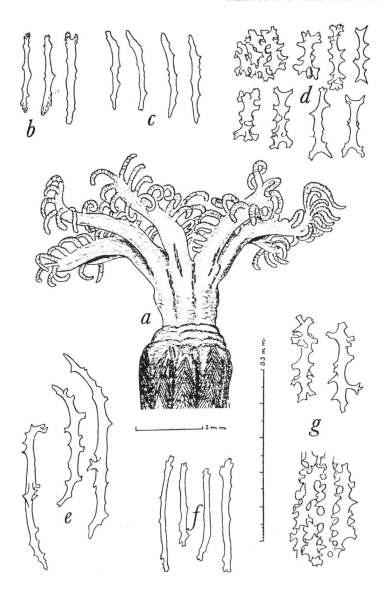

Figure 6.—OCTOCORALLIA. *Telesto riisei. a*, Polyp; *b–g*, sclerites from two different specimens (*b, f*, anthocodial sclerites; *c, e*, nonfusing sclerites from body wall; *d, g*, partially fused sclerites from body wall) (from Bayer, 1961).

Order GORGONACEA

Members of this order, known as gorgonaceans, are characterized by having a skeleton composed of two parts: an outer cortex containing loosely arranged sclerites, and an inner medulla which, when present, has a solid axis of either calcareous or horny material. While there is but one shallow-water representative verified in Hawaii, over 90 deep-water species are recorded (Grigg and Bayer, in press).

The deep-water forms include six species of *Corallium* (family Coralliidae) known as precious or pink corals. Jewelry made from the hard skeleton of *Corallium* is an economic resource for the State of Hawaii (Poh, 1971). However, beds of precious coral must be carefully harvested to avoid overexploitation (Grigg, Bartko, and Brancart, 1973). In addition to pink coral, several genera of the family Primnoidae from deep water are used to make gold coral jewelry, although the most abundant gold coral of the Hawaiian jewelry industry is a species of the zoanthidlike genus *Gerardia* (R.W. Grigg, pers. comm.). In addition, members of the segmented family Isididae, the bamboo corals, are used for jewelry and occur below 100 fathoms.

Gorgonaceans include the sea fans (gorgonians), best known from the tropical Atlantic shallow waters, although many members occur in the western Indo-Pacific region as well. These forms, representing several families of gorgonaceans, are often confused with black corals (order Antipatharia, subclass Zoantharia). However, the skeleton of gorgonians is smooth and the polyps have 8 pinnate tentacles while the skeleton of antipatharians has short spines and the polyps have 6 simple tentacles. The only gorgonian reported from shallow waters in Hawaii is *Acabaria bicolor* in the family Melithaeidae.

Acabaria bicolor (Nutting 1908). This small form (seldom exceeding 40 mm broad by 35 mm high) is branched dichotomously in more than one plane (Fig. 9). The polyps originate from nodes, most often on one side of a branch (Bayer, 1956). Its color is quite variable, from white to yellow (Fig. 8) or pink to red. Specimens are seldom fully exposed, being found under ledges or in rocky crevices at depths from 2 m to 40 m. This species, noted as prevalent on shallow lava flows off the island of Hawaii (R.W. Grigg, pers. comm.), is also known from Oahu, and was first dredged at a depth between 40 and 233 fathoms off Kauai.

126   REEF AND SHORE FAUNA OF HAWAII

FIGURE 7.—(See opposite page for caption).

FIGURE 8.—(See opposite page for caption).

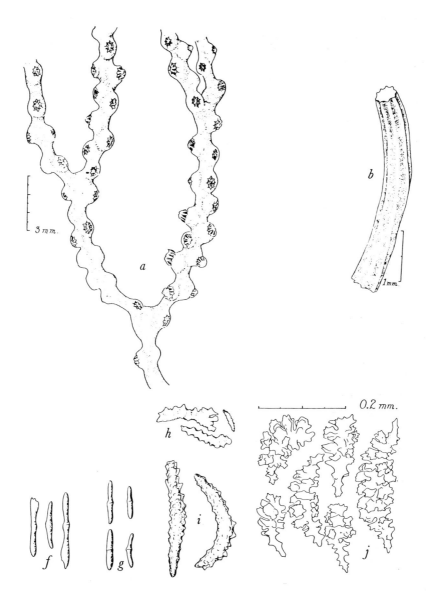

Figure 9.—OCTOCORALLIA. *Acabaria bicolor. a,* Part of a colony (type specimen); *b,* part of axial internode, cortex removed; *f,* sclerites of axis internodes; *g,* of axis nodes; *h,* of tentacles; *i,* of operculum; *j,* of cortex (after Bayer, 1956).

FIGURE 7.—OCTOCORALLIA. *Telesto riisei. a* (top left), Colony with polyps expanded; *b* (right), colony with polyps mostly contracted; *c* (below), branch of colony with polyps expanded (courtesy of Jeff Grovhoug, photos taken at Pearl Harbor).

Figure 8.—OCTOCORALLIA. *Acabaria bicolor.* Colony photographed off Kona, island of Hawaii (courtesy of Chris Newbert).

GLOSSARY (OCTOCORALLIA)

anthocodia: upper, tentacular part of polyp; in many cases can be retracted into anthostele.
anthostele: lower thickened part of polyp's body wall into which anthocodia may be withdrawn; commonly stiffened with sclerites.
axial internodes: axial components, which in the Family Melithaeidae (Order Gorgonacea) are composed of hard inseparably fused calcareous sclerites; internodes are separated by softer, nonconsolidated **axial nodes**.
axial nodes: see **axial internodes**.
axis: central supporting structure of Gorgonacea: it may be composed of sclerites (consolidated or unconsolidated), or horny with more or less scleritic calcareous matter.
base: lower part of colony; often the portion near attachment.
calyx: equivalent to **anthostele**.
capitular: upper part of colony of Alcyonacea, commonly includes the short upper region of the polyps.
cortex: outer layer of tissue of Gorgonacea, including the gastric cavities; in contrast to the axis.
mesoglea: jellylike connective substance containing cells and cellular elements, which occurs between the outer cellular layer (ectodermis) and inner cellular layer (endodermis).
monopodial: pattern of growth in which main stem and all branches of colony are permanently topped by terminal polyps and continue to elongate by means of a growth zone just below each polyp.
spindle: calcareous sclerite formed in mesoglea and occurring as a straight or nearly straight element pointed at both ends.
stolon: basal membranous expansion from which polyps arise.

REFERENCES (OCTOCORALLIA)

Bayer, F. M.
    1952. Descriptions and Redescriptions of the Hawaiian Octocorals Collected by the U.S. Fish Commission Steamer *Albatross*. 1, Alcyonacea, Stolonifera, and Telestacea. *Pacific Science* 6(2): 126–136.
    1956. Descriptions and Redescriptions of the Hawaiian Octocorals Collected by the U.S. Fish Commission Steamer *Albatross*. 2, Gorgonacea: Scleraxonia. *Pacific Science* 10(1): 67–95.
    1961. *The Shallow-Water Octocorallia of the West Indian Region: A Manual for Marine Biologists.* The Hague: Martinus Nijhoff.

Grigg, R. W., B. Bartko, and C. Brancart
    1973. *A New System for the Commercial Harvest of Precious Coral.* UNIHI-Sea Grant-AR-73-01. 6pp., 3 figs.

Grigg, R. W., and F. M. Bayer
    In press. Present Knowledge of the Systematics and Zoogeography of the Order Gorgonacea in Hawaii. *Pacific Science*

Naval Undersea Center
    1974. Pearl Harbor Biological Survey: Final Report (30 Aug. 1974). *NUC TN 1128* (Appendix E).

Nutting, C. C.
    1908. Descriptions of the Alcyonaria Collected by the U.S. Bureau of Fisheries Steamer *Albatross* in the Vicinity of the Hawaiian Islands in 1902. *Proc. U.S. National Mus.* 34(1624): 543–601. Pls. 41–51.

Poh, K-K.
    1971. *Economics and Market Potential of the Precious Coral Industry in Hawaii.* R. W. Grigg (ed.). UNIHI-Sea Grant-AR-71-03. 22 pp.

Tixier-Durivault, A.
    1970. Les Octocoralliaires de Nha-Trang, Viêt-Nam. *Cahiers Pacifique* 14: 115–236.

Utinomi, H.
    1950. Some Xeniid Alcyonarians from Japan and Adjacent Localities. *Publ. Seto Marine Biological Lab.* 1(3): 81–92.
    1958. On Some Octocorals from Deep Waters of Prov. Tosa, Sikoku. *Publ. Seto Marine Biological Lab.* 7(1): 89–110.

Verrill, A. E.
    1928. *Hawaiian Shallow Water Anthozoa.* B. P. Bishop Mus. Bull. 49. Honolulu.

Verseveldt, J.
    1977. Octocorallia from Various Localities in the Pacific Ocean. *Zoologische Verhandelingen* 150: 1–42. 10 pls., 28 figs.

Subclass ZOANTHARIA

ALL MEMBERS of this subclass have polyps with simple or, at most, bifurcate tentacles. They differ further from Subclass Octocorallia in the arrangement of the septa and retractor muscles. Both solitary and colonial forms occur, and a skeleton is present in stony corals (Scleractinia) and black corals (Antipatharia). Representatives of six orders are found in Hawaiian waters. These include the soft-bodied anemones (Corallimorpharia, Actiniaria, Ceriantharia) and zoanthids (Zoanthiniaria), as well as the two coral groups mentioned.

Order CORALLIMORPHARIA

CHARLES E. CUTRESS

University of Puerto Rico

The Corallimorpharia are solitary or colonial anthozoans in which the tentacles are usually radially arranged and the column is smooth. The mesenterial filaments are without ciliated tracts; siphonoglyphs (elongated ciliated grooves in the throat) are very weak or absent; and the marginal sphincter muscle is very weak or absent, with the result that the animal is unable to close rapidly.

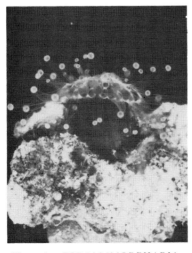

Figure 1.—CORALLIMORPHARIA. *Corynactis* sp.

Family **Corallimorphidae**

Corynactis sp. One undescribed species belonging to the genus *Corynactis* is known from Hawaii. This species is usually colonial, consisting of a large polyp to which several smaller ones are connected by fleshy stolons. Large polyps have both heights and diameters of about 10 mm (Fig. 1). The seventy-odd tentacles are radially arranged and bear acrospheres at the tips. The column is orange brown; the oral disk and tentacles are mostly transparent; and the acrospheres are opaque white to pale orange brown. The species is known from Mauna Lahilahi and Maili Point, Oahu, where it occurs beneath dead coral in waist-deep water.

Order ACTINIARIA

The Actiniaria, commonly known as sea anemones, are solitary anthozoans which usually have cyclically arranged tentacles and various kinds of warts and vesicles on all or part of the column. The mesenteries, after the first twelve, appear as pairs in the exocoels. Mesenterial filaments, with few exceptions, have ciliated tracts. The first major account of Hawaiian actiniarians was by Verrill (1928). In the present treatment 20 taxa are included, taking into account the various taxonomic changes necessitated by more recent study. Of the identified species recorded herein, five are known only from Hawaii, while the majority are more widespread Indo-Pacific forms.

To use the following key to genera of Hawaiian actiniarians adequately, one must not only examine external features but in some cases dissect the animal or make microscopic preparations. The glossary of technical terms, and the specific diagnoses and photographs following the key are provided to aid in the identification of species.

PRESERVATION TECHNIQUES

Anemones to be used for anatomical examination should be properly anesthetized and preserved. Anesthetics found most useful are magnesium chloride (34 percent solution in distilled water), chlorotone (only slightly soluble so a few crystals suffice) and propylene phenoxetol (introduced as an emulsion which is made by shaking two or three drops of phenoxetol with 10 to 20 ml of sea water). Menthol crystals are sometimes useful as a preanesthetic for very sensitive species.

An anemone to be anesthetized should be put into a container of fresh sea water just large enough to allow for full expansion of the animal plus the amount of preservative that will later be added. Some anemones expand in the dark, others in bright light. Allow the anemone to expand fully before introducing the anesthetic. Then add the chemical slowly, a little at a time, in such a way that the specimen will not be disturbed. Anemones that close up during anesthetization rarely open again. Some species will become anesthetized (no reaction to a strong stimulus) in an hour or so; others may require up to 24 hours. Patience is all-important. Anemones with weak musculature, evidenced by slow closure after stimulation, often yield better preserved specimens when the preservative is poured directly over the fully expanded, unanesthetized animal.

Anemones are best preserved by adding formalin to the container in which they were anesthetized. Use 100 percent formalin (37 percent formaldehyde) in sufficient quantity so that the resulting solution will be 10 to 15 percent. Pipette or inject some of the formalin into the coelenteron. Allow the specimen to harden for at least 8 hours before transferring to a storage bottle, otherwise it may collapse. The initial 10 to 15 percent formalin preservation should be with neutral formalin; paraformaldehyde also may be used. For long-term storage, specimens may be transferred to 70 percent undenatured ethanol.

KEY TO GENERA OF HAWAIIAN ACTINIARIA

1	With acontia	12
	Without acontia	2
2(1)	Tentacles with sphincters at their bases	3
	Tentacles without sphincters	4

3(2)	Tentacles very numerous; column smooth	*Boloceroides*
	Tentacles few; lower column with vesicles	*Bunodeopsis*
4(2)	Midcolumn with ring of inflatable outgrowths	*Triactis*
	Column smooth or with vesicles or warts	5
5(4)	Column vermiform; base rounded and inflatable	*Edwardsia*
	Column not vermiform; base flat and adherent	6
6(5)	Tentacles in radial rows	10
	Tentacles in cycles	7
7(6)	Marginal spherules bearing holotrich nematocysts, a type not found in the column	8
	Marginal pseudospherules with same kind of nematocysts as column. Column with rows of adhesive warts	*Actiniogeton*
8(7)	Column smooth	*Anemonia*
	Column with vesicles or warts	9
9(8)	Column with rows of adhesive warts	*Anthopleura*
	Column with crowded, inflatable vesicles	*Cladactella*
10(6)	Few tentacles per endocoel	*Antheopsis*
	Many tentacles per endocoel	11
11(10)	More than one row of tentacles per endocoel	*Stoichactis*
	Tentacles composed of conical marginal and papilliform discals	*Heteranthus*
12(1)	Always symbiotic with pagurid crabs	13
	Not symbiotic with pagurids	14
13(12)	Acontia abundant, with one kind of nematocyst	*Calliactis*
	Acontia sparse, with two kinds of nematocysts	*Anthothoe*
14(12)	Column invested with cuticle and debris	15
	Column smooth and clean	16
15(14)	Tentacles usually blunt to capitate	*Telmatactis*
	Tentacles slender and acuminate	*Epiphellia*
16(14)	Column divided into regions	*Diadumene*
	Column not divided into regions	*Aiptasia*

Family **Boloceroididae**

Boloceroides mcmurrichi (Kwietniewski 1898) [syn. *Nectothela lilae* Verrill 1928]. This soft, flaccid anemone has, in comparison to its small body, an extremely large crown of tentacles (Fig. 2). Specimens with a crown diameter of 15 cm and more than 400 tentacles have been observed, but the majority are less than half this size. The tentacles, of which the innermost are very long, are smooth, adhesive, and readily shed, breaking off just below the sphincter muscle near the attachment to the oral disk. The short column is smooth, thin-walled, and translucent. The base is weakly adherent. The column is pale raw sienna. The oral disk is pale violet or bluish gray except for the colorless directive radius and opaque white radial lines between the tentacles. The tentacles vary from white to pale raw sienna to sepia to reddish brown and may be unmarked or may exhibit transverse bands of pale bluish gray, violet, or opaque white.

This anemone is hermaphroditic and oviparous; it reproduces sexually in the spring and asexually in the fall. The asexual young arise as buds on the outer tentacles just above the tentacle sphincter. These asymmetrical young are shed when they have 10 to 30 tentacles. The anemone swims readily by rhythmic pulsations of its tentacled crown. The large nudibranch *Baeolidia major* closely mimics and preys on *B. mcmurrichi*.

The species is known from Pearl Harbor, Honolulu Harbor, Kaneohe Bay, and Malaekahana, Oahu, where it is found on sand or mud bottoms from low water to about 15 m. At times the anemones are so abundant that they touch one another. The extra-Hawaiian range is S. W. Australia, S. E. Africa, Red Sea, India, and S. Japan.

Bunodeopsis medusoides (Fowler 1888). This anemone (Fig. 3) has a height of 20 mm and a base diameter of 10 mm. The tentacles, numbering about 30, are long and adhesive with pebbled surfaces and sphincter muscles at the bases which make them deciduous. The upper third of the column is smooth, thin-walled, and almost transparent. The lower column is thick-walled and bears inflatable vesicles of which the largest one to six are ovoid and papillate, whereas the numerous small ones are more or less stellate. The weakly adherent base is several times broader than the oral disk. The upper part of the column, the tentacles, and the oral disk are pale greenish brown except for the frosty white tips of the tentacles and an opaque white directive radius on the oral disk. The lower part of the column varies from greenish brown to sepia, with alternating patches of golden brown, violet, and white just below the large vesicles. The large vesicles are green to greenish brown; the small ones are gray to white.

Usually closed in a conical lump during the day, this anemone becomes active at night. It moves by creeping on the ciliated pedal disk, by looping in inchworm fashion, or by swimming by means of rhythmic pulsations of the tentacled crown. When touched, this anemone frequently ejects from the mouth fully formed young. These young are asexually derived from swallowed tentacles which have regenerated in the coelenteron. The species is dioecious and is probably oviparous.

Figure 2.—ACTINIARIA. *Boloceroides mcmurrichi*.

Figure 3.—ACTINIARIA. *Bunodeopsis medusoides*. Painting by Vivian W. Clement.

B. medusoides has been found sparingly about Coconut Island in Kaneohe Bay, Oahu. It is found on dead coral in shallow water. The extra-Hawaiian range is Tahiti and the Tuamotu Islands.

Family **Aliciidae**

Triactis producta Klunzinger 1877 [as *Sagartia pugnax* Verrill 1928 in part]. This is the anemone carried in the claws of the little xanthid crab *Lybia edmondsoni* (Fig. 4). Duerden (1905) described this association but mistakenly called the anemone *Bunodeopsis*. *T. producta* is a soft-bodied anemone about 15 mm in both height and diameter (Fig. 5). Specimens this size have up to 60 long, slender, pebble-surfaced tentacles. The column is tall and smooth except for a ring of up to 24 stalked, compound, inflatable outgrowths borne on a flangelike fold approximately midway between the margin and the base. Young specimens lack these outgrowths. The species is dioecious and viviparous. The anemone is translucent dark sepia except for the frosty white tips of the branches and the hemispherical swellings on outgrowths of the column and, occasionally, at the tips of the tentacles. When deprived of sunlight, as when carried by the secretive crab, these anemones become pale brown or white and often have bright chartreuse tentacles.

In some areas this anemone is found free-living, but in Hawaii it has been collected only with the xanthid crab *Lybia* and, in this association, is an inhabitant of shallow coral reefs. It has been collected from Hanauma Bay, Waikiki Reef, and near Maili Point, Oahu. The extra-Hawaiian range is the Red Sea, India, and E. Australia.

Family **Actiniidae**

Anemonia mutabilis Verrill 1928. This is a soft, flaccid anemone, sticky to the touch, with a crown diameter and height up to 40 mm and 25 mm, respectively (Fig. 6). The column is smooth except for a ring of prominent marginal spherules which may be few or absent in small specimens. The base is broad and weakly ad-

Figure 4.—ACTINIARIA. *Triactis producta,* carried on claws of xanthid crab, *Lybia edmondsoni.*

Figure 5.—ACTINIARIA. *Triactis producta.*

herent. The nonretractile tentacles, numbering to 96 but usually fewer, are smooth, tapered, and acuminate; the innermost tentacles are about equal in length to the radius of the oral disk and the outermost are half this length. The mouth frequently protrudes above the surface of the broad oral disk. The species is dioecious. The color of the column varies from olive green to sepia to orange brown. The marginal spherules are bright orange brown. Except for a reddish brown area around the mouth, the oral disk is sepia with white triangles running from the tentacles about halfway to the mouth. The siphonoglyphs are pale salmon or white. The sepia tentacles are cream proximally and usually have bright pink tips. The base is salmon.

In Hawaii this species is known from Manana Island, Black Point, and Waikiki Reef, Oahu. It occurs chiefly on the undersurfaces of basaltic rocks in shallow water. Specimens occasionally found on reefs are small and pale in color.

Anthopleura nigrescens (Verrill 1928) [syns. *Cladactella obscura* Verrill 1928 and *Tealiopsis nigrescens* Verrill]. Large specimens have crown diameters of 30 mm and heights to 15 mm (Fig. 7). The column bears up to 48 or a few more longitudinal rows of prominent, simple warts to which pieces of shell and rock adhere. Prominent marginal spherules number from 1 to about 48, or may be absent. The base is broad and strongly adherent. The oral disk is about as broad as the base. The actinopharynx (throat) may have one to six siphonoglyphs instead of the normal two. The tentacles, numbering to 96 or a few more, are slender, smooth, and acuminate, the innermost about equal in length to the diameter of the oral disk and the outermost only slightly shorter. The species is dioecious and larviparous, but division by longitudinal fission is commonplace. The anemone is generally a dark violet sepia (almost black), but occasionally brick red specimens are found. The base and lower column are a pale shade of the general color. Each column wart has a central whitish spot, and the marginal spherules are white to yellow orange. The tentacles may all be of the general color, or the outermost may be nearly white, or some may have several whitish spots on their adoral surfaces.

Throughout the Hawaiian Islands, this species is common intertidally in holes or crevices in basaltic rock, sandstone, or limestone. When exposed to the surf, it is

Figure 6.—ACTINIARIA. *Anemonia mutabilis.*

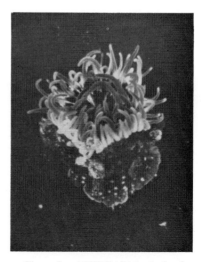

Figure 7.—ACTINIARIA. *Anthopleura nigrescens.*

invariably coated with pieces of shell or rock, but when found in still water or subtidally, it is usually free of debris. A small nudibranch, *Herviella mietta,* mimics and feeds upon *A. nigrescens* (see Rosin, 1969). Dunn (1974c) recently redescribed *A. nigrescens* from Hawaiian waters.

Anthopleura sp. *a* is found at Kaena Point, Kahuku, and on wave-swept rock benches elsewhere on Oahu. It has a crown up to 50 mm in diameter and a height of 20 mm. The column bears about 48 longitudinal rows of large, simple, adhesive warts which at the margin are reduced in size and are close-set on small lobes of the column. At the margin some or all rows of warts terminate with a large, white, marginal spherule. The greenish brown tentacles number to 96 or a few more and are all approximately the same length, equal to the radius of the oral disk. The column is a washed out, pale green. Each wart has a central rusty orange spot.

Anthopleura sp. *b* is known from Pearl Harbor and Kaneohe Bay, Oahu, and has a crown diameter up to 30 mm and a height almost twice that. The column is vase-shaped and bears about 48 longitudinal rows of well-spaced, prominent warts. There are up to 48 conspicuous, white, marginal spherules. The column is grayish, and the warts, greenish brown. The 96 or so tentacles are about the same length and equal to the radius of the oral disk. They are greenish brown, some or all being white proximally.

Actiniogeton sesere (Haddon & Shackleton 1893). Large specimens have crown diameters of 35 mm and heights of 25 mm (Fig. 8). The column bears up to 24 longitudinal rows of prominent, simple warts. Prominent marginal spherules number to 48. The base is broad and strongly adherent. The oral disk is about as broad as the base, and the tentacles are crowded at its periphery. The pharynx may lack siphonoglyphs or have as many as eight. The tentacles, numbering to 96 or a few more, are short, acuminate, and all about the same length. The species is dioecious and individuals frequently undergo longitudinal fission. The anemones are variable in color and markings. Those in sheltered bays have bright green tentacles and oral disks. The latter may be mostly green or be more or less completely covered

Figure 8.—ACTINIARIA. *Actiniogeton sesere.*

Figure 9.—ACTINIARIA. *Cladectella manni.*

with flaky, opaque white around the mouth that extends as radial lines toward the tentacles. The proximal third of the tentacles is diffuse white. The marginal spherules are white and have one to three green spots on their outer surfaces. The column is white near the base, becomes pale pink midway, and is pinkish gray at the margin. The column warts are white and have a bright green spot in their centers. On exposed coasts the anemones have brownish green tentacles and oral disks, with the latter mostly covered by opaque white, and the column is a rosy pink.

In Hawaii this species has been found about Coconut Island in Kaneohe Bay and on Manana Island, Oahu. On exposed coasts the column warts, usually debris-free, may adhere to considerable shell and gravel. The extra-Hawaiian range is N. E. Australia. The name *sesere* should be used advisedly, because specimens from the type locality have not been compared with Hawaiian specimens. (See also Dunn, 1974b.)

Cladactella manni (Verrill 1899). Large specimens have crown diameters of 100 mm and heights of 40 mm (Fig. 9). The column is clean and densely covered with large, simple, nonadhesive vesicles. As many as 96 marginal spherules are more or less concealed in the marginal fold, but these may be few or absent in small specimens. The base is broad and moderately adherent. The tentacles, numbering to 192, are smooth, tapered, acuminate, and all about the same length. The species is dioecious. Numerous young about 1 mm in height have been found in small tide pools near large specimens. The column varies from dark maroon to dark brownish green to dark copper green. The marginal spherules are white. The tentacles are dark rose, becoming greenish proximally. The oral disk is greenish at the periphery, becoming sepia near the mouth. The throat is white to pale pink, and the two siphonoglyphs are crimson, as is the base.

This anemone has been found on the undersurfaces of overhanging sandstone slabs near high water at Kahuku, Oahu, and in tide pools at Kalaupapa, Molokai. This species was first described from Oahu by Verrill (1899) as *Bunodactis manni* and then referred to *Cladactella* in his 1928 paper.

Figure 10.—ACTINIARIA. *Antheopsis papillosa*.

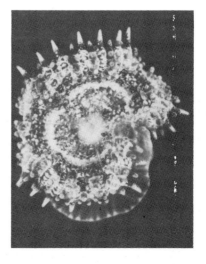

Figure 11.—ACTINIARIA. *Stoichactis* sp.

Family **Stoichactinidae**

Antheopsis papillosa (Kwietniewski 1898) [syns. *Macranthea cookei* Verrill 1928, and *Radianthus papillosa* Kwietniewski, in Dunn, 1974a]. Large, sand-dwelling specimens (Fig. 10) have crown diameters of 150 mm and heights of 200 mm, whereas those living in tide pools are much shorter. The cylindrical to vase-shaped column bears up to 96 longitudinal rows of small adhesive warts which become inconspicuous proximally. At the margin there are 48 small protuberances (pseudospherules). The base is usually about half the diameter of the oral disk and is strongly adherent. The oral disk is broad and usually has five to six temporary folds or undulations at the periphery. The tentacles, numbering to about 384, are distributed on the oral disk from the margin to the edge of the mouth. All are approximately the same length, equal to about half the radius of the oral disk. All tentacles may be smooth, or those of several inner cycles may bear transverse protuberances on their adoral surfaces. Small, extra, endocoelic tentacles may be absent or number from one to three on older endocoels (the space between a pair of mesenteries). The species is dioecious and viviparous, giving birth to young with 8 to 16 tentacles. The base and proximal part of the column are white and are frequently marked with patches and streaks of yellow or orange. Distally the column becomes gray, greenish brown, or violet with white warts. The oral disk is greenish brown or, occasionally, violet and is more or less covered with opaque white. The tentacles are greenish brown, sometimes tinted violet, and the protuberances are opaque white.

This species is known from tide pools at Laie and Kahuku and from sand and mud bottoms in shallow water in Kaneohe Bay, all on Oahu. The extra-Hawaiian range is Java.

Stoichactis sp. A small, flaccid, sticky species of *Stoichactis* (Fig. 11) is occasionally found adhering to the alga *Padina* in Kaneohe Bay, Oahu. Large specimens have crown diameters of 15 mm and heights of 7 mm. The 300 or more short tentacles are radially arranged in up to four rows abreast over the stronger endocoels. The smooth column is pink near the base and brown above. The oral disk and tentacles are translucent brown with varying patterns of opaque white.

Family **Phymanthidae**

Heteranthus verruculatus Klunzinger 1877. Large specimens (Fig. 12) have crown diameters of 12 mm and heights of 7 mm. The cyclically arranged marginal tentacles, numbering to 66 or a few more, are smooth and acuminate, the innermost slightly exceeding in length the diameter of the oral disk, the outermost about half as long. The discal tentacles are papilliform, inconspicuous, and radially arranged, a single row of one to five located over each endocoel. The upper column has longitudinal rows of conspicuous warts which are tentaclelike at the margin. The lower column is essentially smooth. The diameter of the strongly adherent base approximately equals that of the oral disk. The limbus is narrow but distinct. The species is dioecious; however, individuals undergoing longitudinal fission are commonplace. The column changes from greenish brown on the upper part to whitish to pale chartreuse near the base. The warts are bright green, each with a central white spot. The oral disk is dark greenish brown to sepia, overlaid by varying amounts of opaque white, the white sometimes covering all the disk except the mesenterial insertions. Not infrequently the white is distributed as numerous fine flecks and spots. The marginal tentacles are pale greenish brown, either unmarked or with a dense pattern of opaque white flecks and round spots. The discal tentacles are the same color as the oral disk.

In Hawaii this species is known only from Coconut Island in Kaneohe Bay, Oahu, where it occurs on sills of concrete spillways between fish ponds. Here the anemones in the largest population found, about half a square meter, were so dense they touched one another. The extra-Hawaiian range is the Red Sea and E. Australia.

Family **Isophellidae**

Telmatactis decora (Hemprich and Ehrenberg 1834). (See Ehrenberg, 1834) [syns. *Sagartia longa* Verrill 1928, and, in part, *Sagartia pugnax* Verrill 1928]. This distinctive species (Fig. 13) occasionally attains a height of 80 mm and a column diameter of 25 mm, but is usually half this size. Except for a short, clean collar at the margin and a clean area near the base, the column is covered with a thin, wrinkled cuticle heavily invested with fine detritus. The column lacks tenaculi (solid chitin-tipped adhesive papillae) but has a few inconspicuous cinclides (pores in the column wall) on the proximal half. The base approximates the column in diameter and is strongly adherent. The tentacles, maximally 48 but usually fewer, are slightly swollen proximally but taper abruptly to tips that may or may not be capitate. The innermost 12 tentacles approximate in length the diameter of the oral disk; the outermost are about one-fourth as long. The acontia are thin, white, and reluctantly expelled. The species is dioecious, but nothing is known of its development. The column, beneath an ocher cuticle and detritus, is white to pale salmon, becoming pale violet near the margin. The oral disk is colorless to pale violet brown, covered more or less completely with flecks of opaque white. Tentacles of the inner cycle have series of violet brown, opaque white, and translucent transverse bands. Outer tentacles are more or less colorless and without pattern. Occasional specimens have oral disks and tentacles colored a uniform sepia, opaque white, or chartreuse.

Found in many places throughout the Hawaiian Islands, this form occurs in shallow water attached to the undersides of corals or limestone rocks. Young specimens are frequently carried in the chelae of the little xanthid crab *Polydectus cupu-*

Figure 12.—ACTINIARIA. *Heteranthus verruculatus*.

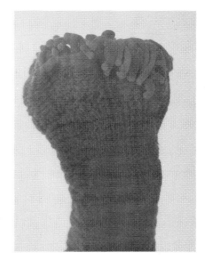

Figure 13.—ACTINIARIA. *Telmatactis decora*.

lifera (Fig. 14). The extra-Hawaiian range is the Red Sea, Zanzibar, India, S. Japan, Christmas Island (Pacific Ocean), and Tahiti.

Epiphellia pusilla (Verrill 1928) [syn. *Sagartia pusilla* Verrill]. Large specimens (Fig. 15) have heights of 15 mm and midcolumn diameters of 5 mm. The tentacles, numbering to 48, are slender, smooth, and acuminate; the length of the outer ones is equal to the diameter of the oral disk, that of the inner ones, twice the diameter. The upper third of the column is thick-walled and has 12 to 18 longitudinal rows of 1 to 10 inconspicuous cinclides through which thin, white acontia are readily expelled. A band of sand adheres to weak tenaculi just below the thin-walled part of the column. The base is about as wide as the column and is weakly adherent. The species is dioecious. The thick-walled parts of the column and base are translucent white. The tentacles are transparent with an opaque white band proximally and with spots and flecks on the adoral surfaces. The thin-walled part of the column and the oral disk are translucent yellow to pale olive green, the disk having an irregular pattern of opaque white.

This species has been collected from Kahala, Oahu, and Nawiliwili Bay, Kauai, where it occurs just below low-tide level attached to the edges and undersurfaces of dead coral which is partially embedded in mud and fine sand.

Epiphellia humilis (Verrill 1928) [syn. *Edwardiella carneola* Verrill 1928, and *Phellia humilis* Verrill]. Large specimens (Fig. 16) have heights of 30 mm and midcolumn diameters of 5 mm. The tentacles, numbering 20 to 30 (usually 24), are smooth and slightly tapered but blunt (not capitate). The innermost tentacles are about equal in length to the diameter of the oral disk; the outermost ones are about half as long. The body is divisible into a short, clean, thin-walled upper portion; a long, thick-walled middle portion heavily invested with sand grains and other detritus; and a short, clean, thin-walled, inflatable but adhesive base. The thick-walled portion of the column bears about 12 longitudinal rows of inconspicuous cinclides and numerous, scattered, inconspicuous tenaculi. Acontia are few, slender, and white. The species is dioecious. The column is translucent but is covered, except for

Figure 14.—ACTINIARIA. *Telmatactis decora* carried on claws of crab, *Polydectus cupulifera*.

Figure 15.—ACTINIARIA. *Epiphellia pusilla*.

a short region at the margin and base, with black sand distally and orange to ocher detritus proximally. Occasionally, the black sand is not present, and the invested part of the column is orange to ocher throughout. The oral disk is white to cream with sepia radial lines marking the mesenterial insertions. The tentacles are translucent, opaque white distally and proximally, and with a pattern of several sepia-colored W's and V's adorally.

This species is known from Kahala Reef, Oahu, and Nawiliwili Bay, Kauai, where it occurs on inner reefs, usually attached to the undersurfaces of dead coral or rocks. It is sometimes unattached in the sand.

Family **Hormathiidae**

Calliactis polypus (Forskål 1775) [syn. *Calliactis armillatus* Verrill 1928]. In Hawaii, this anemone is known only in association with pagurid (hermit) crabs. The largest specimens have, in full extension, both maximum heights and base diameters of 80 mm, but most are about half this size. Except for a short, thin-walled, cuticle-free collar near the margin, the column is thick, firm, and covered with a thin cuticle. Three horizontal rows of cinclides borne on conspicuous, white protuberances occur a short distance above the edge of the base; when the animal contracts, the three rows merge into one. Normally, the base is broad and firmly attached to the various species of gastropod shells (Fig. 17) worn by different pagurid crabs belonging to the genus *Dardanus*. The tentacles, numbering to about 550 in large specimens, are slender, smooth, and acuminate, the innermost approximately equal in length to half the radius of the oral disk, and the outermost about half this length. The profuse, salmon-colored acontia, containing only one kind of nematocyst, are readily expelled. The species is dioecious and oviparous, the eggs hatching into typical planulae. The column varies from ocher to gray to dirty orange and is overlaid more or less with patches and/or broad streaks of plum red. Near the base it is marked with short, thin, longitudinal white lines. The tentacles are usually translucent, pale brownish to grayish, and are marked adorally with narrow transverse lines of pale rose and

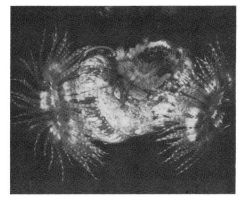

Figure 16.—ACTINIARIA. *Epiphellia humilis*.

Figure 17.—ACTINIARIA. *Calliactis polypus* (large ones at each end of shell) *Anthothoe* sp. (on columella and lip of shell).

opaque white. Specimens from off shore frequently have unmarked, bright yellow, orange, or salmon pink tentacles.

In Hawaii this species is common on and around reefs to depths of about 40 meters. In addition to the pagurid crabs (*Dardanus*) with which it is always found, other animals are commonly associated. Attached to the columella of the gastropod shell is an undescribed anemone, *Anthothoe* sp. (Fig. 17), and within the umbilicus dwells the polyclad *Stylochoplana inquilina*. Roaming over and attached beneath the base of *Calliactis* are, respectively, the amphipod *Elasmopus calliactis* and the barnacle *Koleolepas tinkeri*. Elsewhere in its range, *C. polypus* occurs with other crabs and other symbionts. The extra-Hawaiian range is the Red Sea, E. Africa, India, Australia, French Polynesia, S. Japan, and W. Central America.

Family **Sagartiidae**

Anthothoe sp. Large expanded specimens (Fig. 17) have both heights and crown diameters of 15 mm and a base diameter of 30 mm. The column is smooth and clean, with a few inconspicuous cinclides that are frequently difficult to discern. The column flares broadly at the base where it is irregular in outline and much convoluted. A full complement of tentacles in large specimens is 192, but the usual number is a few more than 96. The tentacles are smooth, tapered, and acute with a tendency to be retroflexed. The innermost are about equal in length to the radius of the oral disk; the outermost are about half this length. Frequently, some or all of the innermost 12 tentacles appear thicker and more opaque than the others. At such times they have a type of large nematocyst (holotrich) not found elsewhere in the anemone. The sparse, thin acontia are white and contain two kinds of nematocysts. The species is dioecious, but asexual reproduction by basal laceration is commonplace. The column is a translucent cream color marked with irregularly dispersed and interrupted fine streaks of yellow ocher which become more numerous near the base. The oral disk is translucent cream with flecks of opaque white around the mouth. The tentacles are translucent cream, occasionally with a few diffuse transverse bars of pale greenish brown. When contracted and when preserved, the tentacles have gray cores.

This species occurs with *Calliactis polypus* and its associates. *C. polypus* is found attached to almost any part of the gastropod shell (in Hawaii usually *Tonna*

Figure 18.—ACTINIARIA. *Aiptasia pulchella*, large form.

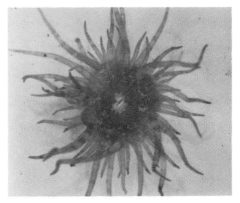

Figure 19.—ACTINIARIA. *Aiptasia pulchella*, small form (oral disk).

or *Turbo)*, but *Anthothoe* is almost always found on the columella and along the inner lip beneath the body of the hermit crab. The range of the species is the same as that of *C. polypus*.

Family **Aiptasiidae**

Aiptasia pulchella Carlgren 1943 [syn. *A. californica* Carlgren 1952]. Large specimens (Fig. 18) of this soft, flaccid anemone have both heights and crown diameters of 80 mm and midcolumn diameters of 12 mm. The tall, vase-shaped column is thin-walled, smooth, and clean. Twelve longitudinal rows of one to four prominent cinclides band the midcolumn. The base is about the same diameter as the oral disk and is weakly adherent. The tentacles number to about 120, the innermost equal in length to twice the diameter of the oral disk, the outermost about half as long. These are slender, smooth, evenly tapered, acute, and highly contractile. The profuse, long, white acontia are readily expelled. The species is dioecious and viviparous. The anemone is usually a dark, rich brown, but specimens living in dark places are pale brown or even white. Markings may be absent, or there are sometimes small blotches of white on the column near the margin, on the oral disk, and near the attachment of the inner tentacles. Occasionally, the directive radius on the oral disk is white.

The species occurs in two distinct forms which, at first sight, may be taken for different species. The small form, one-fourth or less the size of the above-described form, is pale greenish brown and is profusely marked all over with white flecks (Fig. 19). The two forms do not coexist in an area, but one may supplant the other. Both are common throughout the Hawaiian Islands in shallow water on reefs and in bays. The extra-Hawaiian range is S. Japan, French Polynesia, W. Central America, and the Gulf of California.

Family **Diadumenidae**

Diadumene leucolena (Verrill 1866). Large specimens (Fig. 20) have column diameters of 10 mm and heights of 40 mm. The column is distinctly divided into two regions by a collarlike fold a short distance below the margin; it is soft and smooth except for numerous, prominent cinclides on the proximal two-thirds. The base is a little wider than the column and is firmly adherent. The tentacles, numbering to 96

Figure 20.—ACTINIARIA. *Diadumene leucolena*. Photo by Peter E. Pickens.

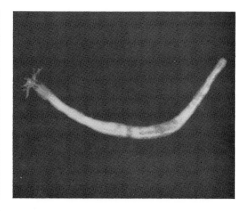

Figure 21.—ACTINIARIA. *Edwardsia* sp. a.

or a few more, are smooth and acuminate, the innermost slightly exceeding in length the diameter of the oral disk and the outermost being about one-fourth this length. Frequently, one or more or all of the inner 12 tentacles (catch tentacles) are thicker and more opaque than the others and are greatly extensible. The profuse, thin, white acontia are readily expelled. The species is dioecious and oviparous. The over-all color of this anemone varies from a dark, dirty green to greenish brown to dirty white. The column bears lighter colored, longitudinal lines which correspond to the mesenterial insertions. Occasionally, the directive radius on the oral disk is lighter than the general color.

This species is most likely an accidental introduction in Hawaii, probably having been brought in with oysters, upon which it is commonly found. It is known from Pearl Harbor, Honolulu Harbor, and the Ala Wai Canal, Oahu. The extra-Hawaiian range is E. United States, West Indies, and California.

Family **Edwardsidae**

Edwardsia sp. *a*. This species (Fig. 21) is up to 60 mm in height and 3 mm in diameter and has 14 to 18 (usually 16) long, slender tentacles. Except for a short, opaque white, marginal collar and a translucent physa, the column is covered with a thick, shaggy, rusty orange cuticle.

This is a sand-dwelling species found commonly in shallow water near shore at Waikiki, Maili, Kahala, and Coconut Island, Oahu.

Edwardsia sp. *b*. This second, distinct species of *Edwardsia* (Fig. 22) is up to 70 mm in height and 10 mm in diameter. It has 14 to 20 (mostly 16) tentacles which are markedly shorter than those of species *a*. The column is covered with a very thin, ocher cuticle to which fine sand adheres. The clean marginal collar is violet sepia, and the tentacles are densely marked with spots of opaque white.

Like sp. *a*, sp. *b* is sand-dwelling and occurs in shallow water near shore at Waikiki, Maili, Kahala, and Coconut Island, Oahu.

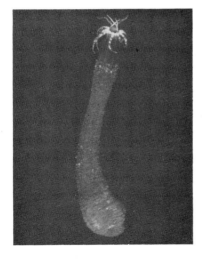

Figure 22.—ACTINIARIA. *Edwardsia* sp. *b*.

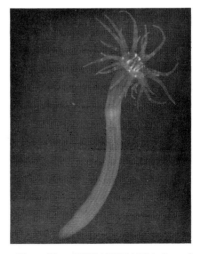

Figure 23.—CERIANTHARIA. *Isarachnanthus bandanensis*.

Order CERIANTHARIA

These are solitary, elongate, tube-dwelling, hermaphroditic anthozoans with long tentacles arranged in two distinct cycles. New mesenteries, after the first six, arise on one side of the body (the multiplication compartment). Only one species, a member of the family Acontiferidae, is known from Hawaiian waters.

Isarachnanthus bandanensis Carlgren 1924 [as *Cerianthus* sp. in Edmondson, 1946]. This "sea anemone" (Fig. 23) has a height and crown diameter up to 100 mm and 60 mm, respectively, and dwells in a tube composed of nematocysts and mucus. The tentacles, numbering maximally 56 but more commonly 48 or 52, occur in two distinct cycles, marginal and labial, the latter half the length of the former. They are long, thin, and acuminate. Although normally smooth, they exhibit several annular thickenings or ridges when partially deflated. The expanded animal characteristically holds the marginal tentacles horizontal and the labials vertical to the surface of the oral disk. The oral disk, approximately the same diameter as the body, is funnel-shaped, tapering into the pharynx. The body is long and smooth and is gradually tapered to the aboral terminus where there is a pore. Although the body wall is thick, it is translucent, revealing the mesenterial insertions as longitudinal lines. The species is oviparous and, like all Ceriantharia, hermaphroditic. The animal is over-all translucent, reddish brown (almost copper), being darkest below the tentacles and palest on the tentacles, oral disk, and aboral end. The only markings of note are opaque white triangles at the bases of the marginal tentacles, the white occasionally extending to the labials.

In Hawaii, the species is known from Ahu o Laka islet in Kaneohe Bay, Oahu, where it lives buried in stable sand. The largest populations, in some places having a density of more than a dozen per square meter, are found at about the zero tide level. The extra-Hawaiian range is Indonesia, French Polynesia, and the Marshall Islands.

GLOSSARY
(CORALLIMORPHARIA, ACTINIARIA, CERIANTHARIA)

acontia: thin threads in certain actiniarians which hang free from the mesenterial edge just below the filament and which contain extraordinary numbers of nematocysts of one or more kinds.

acrosphere: the globular ends of certain tentacles laden with numerous, large nematocysts.

actinopharynx (also **pharynx, throat, stomodaeum**): the tube which leads from the mouth to the coelenteron.

basal laceration (also **pedal laceration**): asexual production of individuals from fragments autotomized from the edge of the pedal disk.

base (also **pedal disk**): the aboral end of a polyp; usually that part attached to the substrate.

catch-tentacles: tentacles of certain actiniarians, usually some or all of the first or second cycles, which are capable of extreme elongation and which contain kinds of large nematocysts not found elsewhere in members of the species.

ciliated tract: see filament.

cinclides: definite small apertures (or weak spots which will rupture readily) in the column wall of certain actiniarians.

coelenteron: the gastrovascular cavity of a coelenterate.

collar (also **marginal fold** or **parapet**): a distinct fold of the column just below the edge of the oral disk; the groove thereby produced is called the "fosse."

column: the body; usually that part of an actiniarian between the margin and the pedal disk.

crown: the entire expanse of tentacles.

cuticle: a dead, noncellular organic layer secreted by the external epithelium.

directive axis: the line passing between the partners of the two pairs of mesenteries having retractor muscles on their exocoelic sides; usually recognizable as the elongate axis of the mouth.

discal tentacles: see **tentacles.**

endocoel: see **mesenteries.**

exocoel: see **mesenteries.**

filament: the thickened free border of a mesentery consisting of three cords or tracts on the upper part and one on the lower. The lateral cords are called "ciliated tracts"; the medial is the "cnidoglandular tract."

holotrich: see **nematocysts.**

labial tentacles: in cerianthids, those tentacles belonging to the cycle nearest the mouth.

marginal fold: see **collar.**

marginal sphincter: a concentration of entodermal circular muscle fibers in the region of the collar or margin. The fibers may be borne on mesogloeal pleats extending into the entoderm in which case the sphincter is said to be "entodermal"; or, should the fibers line cavities within the mesogloea, the sphincter is said to be "mesogloeal."

marginal pseudospherules: see **marginal spherules.**

marginal spherules: vesicles situated on the edge of the collar, or in the fosse, and containing holotrich nematocysts as well as other kinds. "Pseudospherules" resemble the spherules but lack holotrichs.

mesenteries (also **septa**): the thin, usually paired sheets of tissue which partition the coelenteron of Anthozoa. "Perfect" mesenteries are attached to (an insertion on) the pharynx, oral disk, column and base while "imperfect" ones are not or only partially attached to the pharynx. The space between the members of a pair of mesenteries is called the "endocoel" while the space between two pairs is the "exocoel."

multiplication compartment: in cerianthids, the area of the coelenteron opposite the siphonoglyph where new mesenteries are formed.

nematocysts (also **cnidae**): the nonliving, secreted capsules (stinging cells) characteristic of coelenterates. The capsules contain a tube which is continuous with the capsule wall at one end and also a liquid component composed of poison and pain-producing substances. The tube may or may not have a uniform diameter and may or may not bear spines that may or may not be of uniform size. One kind ("holotrichs") have a uniform diameter tube with uniform-sized spines throughout its length.

oral disk: the adoral surface of a polyp; that part between the mouth and the margin over which, to a greater or lesser extent, issue the tentacles.

pseudospherules: see **marginal spherules.**

physa: the inflatable, usually nonadherent, aboral end of certain burrowing actiniarians.

siphonoglyphs: histologically differentiated grooves extending from the corners of the mouth to the inner end of the pharynx and sometimes beyond; usually associated with the directives.

stolons: tubular outgrowths from the base of a polyp from which new individuals may grow.

tenaculi: solid, chitinized papillae on the columns of certain actiniarians.

tentacles: hollow outgrowths from the oral disk; usually one communicates with each mesenterial endocoel and one with each exocoel. In certain anemones, two distinct kinds of tentacles are present, the normal cyclically arranged ones at the periphery of the oral disk, called "marginal tentacles," and papilliform or branched, radially arranged ones on the inner oral disk, called "discal tentacles."

verrucae: ampullaceous adhesive evaginations of the columns of certain actiniarians; usually simple but occasionally compound. The ectoderm in the center of the evagination consists exclusively of tall cells containing adhesive granules.

vesicles: ampullaceous nonadhesive evaginations of the columns of certain actiniarians; either simple or compound. The ectoderm resembles that of the column and contains nematocysts.

warts: see **verrucae.**

REFERENCES
(CORALLIMORPHARIA, ACTINIARIA, CERIANTHARIA)

Carlgren, O. H.
- 1924. Papers from Dr. Th. Mortensen's Pacific Expedition 1914–1916. XVI. Ceriantharia. *Videnskabelige Meddelelser Dansk Naturhistorisk Forening Kobenhavn* 75:169–195. 16 figs.
- 1949. A Survey of the Ptychodactiaria, Corallimorpharia and Actiniaria. *Kungliga Svenska Vetenskapsakademiens Handl.* (Ser. 4) 1(1): 1–121. 4 pls.
- 1950. A Revision of Some Actiniaria Described by A. E. Verrill. *J. Washington Acad. Sciences* 40(1): 22–28. 1 fig.

Duerden, J. E.
- 1905. On the Habits and Reactions of Crabs Bearing Actinians in Their Claws. *Proc. Zoological Soc. London* 2(2): 494–511. Figs. 72–76.

Dunn, D. F.
- 1974a. *Radianthus papillosa* (Coelenterata, Actiniaria) Redescribed from Hawaii. *Pacific Science* 28(2): 171–179.
- 1974b. *Actiniogeton sesere* (Coelenterata, Actiniaria) in Hawaii. *Pacific Science* 28(2): 181–188.
- 1974c. Redescription of *Anthopleura nigrescens* (Coelenterata, Actiniaria) from Hawaii. *Pacific Science* 28(4): 377–382.

Edmondson, C. H.
- 1946. *Reef and Shore Fauna of Hawaii*. B. P. Bishop Mus. Spec. Publ. 22. Honolulu.

Ehrenberg, C. G.
- 1834. *Uber die Natur und Bildung der Coralleninseln und Corallenbänke im rothen Meere*. Abhandlung Königlichen Akad. Wissenschaften (for 1832). Berlin.

Josephson, R. K., and S. C. March
- 1966. The Swimming Performance of the Sea-Anemone *Boloceroides*. *J. Experimental Biology* 44: 493–506.

Rosin, R.
- 1969. Escape Response of the Sea Anemone *Anthopleura nigrescens* (Verrill) to Its Predatory Eolid Nudibranch *Herviella* Baba Spec. Nov. *Veliger* 12: 74–77.

Ross, D. M.
- 1970. The Commensal Association of *Calliactis polypus* and the Hermit Crab *Dardanus gemmatus* in Hawaii. *Canadian J. Zoology* 48: 351–357.

Stevenson, R. A.
- 1963. Behavior of the Pomacentrid Reef Fish *Dascyllus albisella* Gill in Relation to the Anemone *Marcanthia cookei*. *Copeia* No. 4, pp. 612–614.

Uchida, T.
- 1938. Tropical Actinian Forms in Japan (I). *Annotationes Zoologicae Japonenses* 17(3,4): 623–635. 11 figs.

Verrill, A. E.
- 1899. Descriptions of Imperfectly Known and New Actinians: With Critical Notes on Other Species. *American J. Science* 7(art. 21): 205–218. Figs. 22–32.
- 1928. *Hawaiian Shallow Water Anthozoa*. B. P. Bishop Mus. Bull. 49, Honolulu. 3 figs, 5 pls.

ORDER ZOANTHINIARIA

GERALD E. WALSH and RALPH L. BOWERS

University of Hawaii

THE ZOANTHINIARIA (Zoanthidea) is a small group of sessile (attached) and mostly colonial anemonelike anthozoans. In shallow water these organisms, sometimes referred to as soft corals, may form extensive colonies that encrust hard surfaces or live buried in sand up to the level of their oral disks. Although skeletons are not formed, sand grains are found embedded in the tissues of some species, giving support and firmness to the polyps (Edmondson, 1946). The polyps arise from platelike (lamellate) or branching (stolonate) extensions of the body wall called the coenenchyme. This tissue contains gastrodermal canals continuous with the digestive (gastrovascular) cavity of each polyp in the colony. New polyps are budded from extensions of the gastrodermal canals rather than from old polyps. Numerous mesenteries (septa) are arranged in a pattern unlike other anthozoan orders. Muscle tissues borne on the mesenteries produce rather sluggish column and weak retractile movements (Bayer and Owre, 1968).

One shallow-water family (Zoanthidae) is represented in Hawaii by seven species in three genera—*Isaurus, Palythoa,* and *Zoanthus* (Walsh and Bowers, 1971). There is extensive intraspecific variation of the external morphology in *Palythoa* and *Zoanthus* in relation to habitat, and it is necessary to examine microscopic sections of the column for proper identification. For all three genera, the cnidom (description of types, sizes, and distribution of nematocysts) is of great help in identifying species. For this technical information, see Walsh and Bowers (1971). Two of the Hawaiian zoanthids have been studied in connection with their feeding responses (Reimer, 1971). A chemical component (palytoxin) from Hawaiian zoanthids has demonstrated antitumor activity in mice (Quinn and others, 1974).

Isaurus. These forms are without encrustation of the body wall and have a single mesogleal sphincter.

Isaurus elongatus Verrill 1928. Adult polyps—70 mm long, 9 mm in diameter—are usually separate but may cluster in groups of up to 50 individuals. Occasionally, colonies of up to 5 polyps are united by lamellate coenenchyme. New individuals arise as buds from the coenenchyme either near the bases of older polyps or at the ends of short stolons. After a period of development, the young polyp often loses its connection to the parent and stands alone. The column may stand vertically, horizontally, or at an angle; it may be straight or curved. Uniformity of the colony has never been observed with respect to column orientation (Fig. 1). Several rows of tubercles are seen on the upper third of the column of many large polyps. The thin-walled scapus has mesenterial insertions showing through in living specimens. The color in life is light tan (small polyps) to brown with white spots scattered randomly

Figure 1.—ZOANTHINIARIA. *Isaurus elongatus* Verrill 1928 (from Plate 1, Walsh and Bowers, 1971).

(large polyps). Often bright green areas are seen in the upper third of the column. Preserved specimens are brown. There are as many as 50 tentacles and mesenteries and up to 15 capitular ridges. The sphincter has numerous meshes; the mesoglea, numerous gastrodermal canals, cell islets, and lacunae. There are 3 to 5 basal canals

Figure 2.—ZOANTHINIARIA. *Palythoa vestitus* (Verrill 1928) (from Plate 2, Walsh and Bowers, 1971).

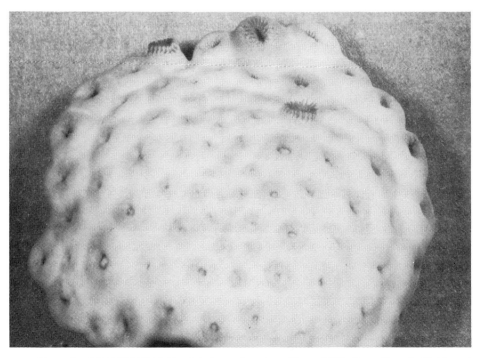

Figure 3.—ZOANTHINIARIA. *Palythoa tuberculosa* (Esper 1791) (from Plate 5, Walsh and Bowers, 1971).

along the entire length of the mesenteries. The mesoglea has numerous gastrodermal canals in the oral third of the column. This species is fairly common in the intertidal zone, where it is usually found on the undersides of rocks in subtidal shallow water, and in crevices on coral reefs. It is known from all the Hawaiian Islands.

Palythoa. In these forms, the body wall is heavily encrusted, and there is a single mesogleal sphincter.

Palythoa vestitus (Verrill 1928) [syn. *Zoanthus vestitus* Verrill]. The polyp is cylindrical when expanded, club-shaped when contracted (Fig. 2). The lamellate coenenchyme is thick, and the animal is heavily encrusted with sand. In the intertidal zone, expanded specimens may be 9 mm high with a diameter of 5 mm. In the subtidal zone, the height may be 14 mm; the diameter, 8 mm. In surge pools the height may approach 26 mm; the diameter, 11 mm. Capitular ridges number up to 30; mesenteries and tentacles, 60. The tentacles are short. The mesoglea has ectodermal, but no gastrodermal, canals. The outer half of the mesoglea of the column is embedded with sand. The inner half has numerous lacunae which contain zooxanthellae. The sphincter is moderately developed, with the largest cavities located distally. There are up to 9 canals at the basal half of the complete mesenteries. The oral disk is green or brown. The species is common on reefs, in surge pools, and in the intertidal and subtidal zones along rocky shores of Oahu, Kauai, Hawaii, and Maui. It has also been found on the reefs in Pago Pago Harbor on the island of Tutuila, American Samoa, and at Atimaono Pass, Tahiti.

Palythoa tuberculosa (Esper 1791). The polyps are immersed in coenenchyme, which may be as thick as 26 mm (Fig. 3). The mesoglea, which contains pigment

Figure 4.—ZOANTHINIARIA. *Palythoa psammophilia* Walsh and Bowers 1971 (from Plate 3, Walsh and Bowers, 1971).

granules, fills the entire space between the polyps. The diameter of living, contracted polyps removed from the colony is as much as 8 mm. There are up to 35 capitular ridges and up to 50 tentacles. The colony is heavily encrusted, the color of the animal being that of the encrusting material. There are numerous gastrodermal canals in the mesoglea. The sphincter is moderately developed. The complete mesenteries have 1 basal canal and up to 5 canals proximal to the filament. Colonies are found in surge pools, along rocky coasts, and on the reefs of Oahu, Kauai, Hawaii, and Maui. This species is known throughout the Indo-Pacific region.

Palythoa psammophilia Walsh and Bowers 1971. The polyps, buried in sand to the level of the oral disks, are erect and cylindrical. Stolonate coenenchyme attaches the colony to coral rubble below the surface of the sand. The epidermis is heavily encrusted with sand. Living polyps are as high as 22 mm and may have a diameter of 9 mm (Fig. 4). Capitular ridges number up to 30; mesenteries and tentacles, up to 60. The tentacles are long. The outer third of the mesoglea is heavily embedded with sand; it has numerous cell islets, ectodermal but no gastrodermal canals, and numerous lacunae with holotrichs and zooxanthellae. The sphincter is weak. There is a dense fibrillar tissue layer at the base of the epidermis of the oral disk. The color of the oral disk is green to light brown. The complete mesenteries have up to 6 basal canals and numerous pigment granules. The colonies are common on the sand flats of Kaneohe Bay, Oahu, the only recorded locality.

Palythoa toxica Walsh and Bowers 1971. The polyp is erect and cylindrical, with a slight swelling at the base, and is heavily encrusted (Fig. 5). The columns of expanded, preserved specimens are as tall as 14 mm with diameters up to 7 mm. The coenenchyme is lamellate. The capitular ridges number up to 30; mesenteries

Figure 5.—ZOANTHINIARIA. *Palythoa toxica* Walsh and Bowers 1971. Disk diameter approximately 7 mm.

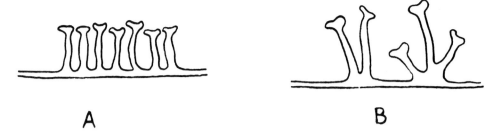

Figure 6.—ZOANTHINIARIA. *a*, Reef growth form of *Zoanthus pacificus* showing crowded polyps with separate bases; *b*, surge pool growth form of *Z. pacificus* showing less crowded polyps with bases united.

and tentacles, 60. Both ectodermal and gastrodermal canals are absent from the mesoglea, which contains pigment granules and is heavily encrusted throughout. The sphincter is weak, with cavities approximately the same size throughout. There are as many as 5 basal canals in the complete mesenteries, some of which extend into the mesoglea of the column. The oral disk is light or dark brown, often with a random pattern of small white spots around the mouth. These colonies have been reported only from surge pools at the Lanai Lookout and Blowhole, Oahu, and from the Hana District, Maui.

The name *toxica* was chosen for this species because a strong toxin is present in the mucus of the gastrovascular cavity. Great care must be taken when handling

Figure 7.—ZOANTHINIARIA. *Zoanthus pacificus* Walsh and Bowers 1971. Oral disk diameter approximately 7 mm.

these polyps. A student collector, who inadvertently touched a colony to an open lesion, required hospitalization for two days (Walsh and Bowers, 1971). The toxic property of this zoanthid species was known to the early Hawaiians who called it *limu make o Hana* (deadly seaweed of Hana) and would smear it on spear points to make them fatal (Moore and Scheuer, 1971). A study on the nature of the toxin from several species of *Palythoa* has been reported (Moore and others, 1975).

Zoanthus. These zoanthids are without encrustation of the body wall and have a double mesogleal sphincter.

Zoanthus pacificus Walsh and Bowers 1971 [syns. *Zoanthus confertus* Verrill 1928, and *Zoanthus nitidus* (Verrill 1928)]. The external morphology of this species varies according to the habitat. On reefs, the polyps grow from lamellate coenenchyme and are in very close proximity to each other, but the bases are separate (Fig. 6a); in surge pools, lamellate coenenchyme gives rise to polyps not as crowded, the bases of which are often united (Fig. 6b); on rocky, wave-washed shores, the coenenchyme is either lamellate or stolonate, and the polyps are single or in groups of two to three, often between polyps of *Palythoa vestitus*. The living, expanded polyp may be 15 mm high; the diameter, 7 mm (Figs. 7, 8). Mesenteries and tentacles number up to 60. The proximal portion of the sphincter has cavities of irregular shape. Pigment granules are found in the epidermis and in the outer half of the mesoglea which has numerous ectodermal and gastrodermal canals. There is a single basal canal along the entire length of the complete mesenteries. The colonies are common on the shores of Oahu, Kauai, Hawaii, and Maui. This species is also known from Pago Pago Harbor on the island of Tutuila, American Samoa, and on the reef at Atimaono Pass, Tahiti.

Figure 8.—ZOANTHINIARIA. *Zoanthus pacificus* Walsh and Bowers 1971. Oral disk diameter approximately 7 mm.

Figure 9.—ZOANTHINIARIA. *Zoanthus kealakekuaensis* Walsh and Bowers 1971. Disk diameter approximately 5 mm.

Figure 10.—ZOANTHINIARIA. *Zoanthus kealakekuaensis* Walsh and Bowers 1971. Disk diameter approximately 5 mm.

Zoanthus kealakekuaensis Walsh and Bowers 1971. The lamellate coenenchyme is highly developed, often extending to half the column height. Polyp bases are separate, and the height of living, expanded polyps may reach 10 mm; the diameter, 5 mm (Fig. 9). Mesenteries and tentacles number up to 54 (Fig. 10). All the cavities of the proximal portion of the sphincter are elongate. The mesoglea is without gastrodermal canals and has no pigment granules. The basal canals of the complete mesenteries are 5 in number at the base of the column. There is a single basal canal at the oral end. These colonies have been reported only from the intertidal zone of Kealakekua Bay, Hawaii, living on lava rock. As in *Palythoa toxica*, the gastrovascular mucus of *Z. kealakekuaensis* appears to contain a toxin and should be handled with great care (Walsh and Bowers, 1971).

GLOSSARY (ZOANTHINIARIA)

basal canals: small canals, located where the mesenteries are attached to the column and running parallel to the mesenteries.

capitulum: upper, short, thin-walled region of the column; sometimes having ridges.

capitular ridges: see **capitulum.**

cell islets: small groups of pigmented cells irregularly scattered in the mesoglea. The islets appear to be groups of ordinary mesogleal cells.

coenenchyme: an extension of the columns (body walls) of polyps; a common tissue which connects the polyps of a colony; it may appear as stolonate, lamellate, or as flattened expansions, which may fill the spaces between polyps.

column: the body wall of the polyp.

encrustations: except for *Zoanthus* and *Isaurus,* zoanthids incorporate sand grains, sponge spicules, foraminifera tests, and other hard particles into the body wall.

filament: see **mesenterial filament.**

gastrodermal canals: canals that connect the gastrovascular cavity of all polyps in the colony. These are found throughout the coenenchyme and can initiate the formation of new polyps.

gastrovascular cavity: the main area for digestion in the polyp, beginning at the mouth (located on the oral disk) and containing various regions of cell specialization, the mesenteries, and the gastrodermal canals.

holotrich: nematocyst with a tube of uniform diameter which is uniformly armored throughout its length; it is typically long, with a thin capsule and a nearly or completely coiled content.

lacunae: open spaces of various sizes in the mesoglea.

mesenterial (septal) filament: the free edge of each mesentery below the pharynx.

mesentery (septum): a longitudinal partition in the gastrovascular cavity of anthozoans which divides the cavity into chambers. The mesenteries are composed of gastrodermis and mesoglea. Complete mesenteries are those continuous from the column wall to the gastrovascular cavity.

mesoglea: a gelatinous transparent matrix that comprises the central portion of the body wall between ectoderm and entoderm.

mesogleal sphincter: present in the capitular region of the column of all shallow-water Hawaiian zoanthid genera; it is composed of a single band of muscle tissue in *Isaurus* and *Palythoa,* a double band in *Zoanthus.*

nematocyst (stinging cells, nettle cells): a cell organoid which consists essentially of a capsule containing a coiled capillary tube which, upon stimulation, discharges to the outside by turning inside out.

oral disk: the tissue that extends between the base of the tentacles and the mouth.

scapus: lower, thick-walled region of the column.

sphincter: see **mesogleal sphincter.**

ectodermal canals: large ectodermally lined canals which penetrate the mesoglea; the canals course in a radial direction and may pass into the mesenteries.

zooxanthellae: vegetative state of symbiotic dinoflagellate algae found in the tissue of some zoanthids.

REFERENCES (ZOANTHINIARIA)

Bayer, F. M., and H. B. Owre
    1968. *The Free-Living Lower Invertebrates.* New York: Macmillan.

Edmondson, C. H.
    1946. *Reef and Shore Fauna of Hawaii.* B. P. Bishop Mus. Spec. Publ. 22. Honolulu.

Moore, R. E., and P. J. Scheuer
    1971. Palytoxin: A New Marine Toxin from a Coelenterate. *Science* 172: 495–498.

Moore, R. E., R. F. Dietrich, B. Hatton, T. Higa, and P. J. Scheuer
    1975. The Nature of the 263 Chromophore in the Palytoxins. *J. Organic Chemistry* 40(4): 540–542.

Quinn, R. J., M. Kashiwagi, R. E. Moore, and T. R. Norton
    1974. Anticancer Activity of Zoanthids and the Associated Toxin, Palytoxin, against Ehrlich Ascites Tumor and P-388 Lymphocytic Leukemia in Mice. *J. Pharmaceutical Sciences* 63(2): 257–260.

Reimer, A. A.
    1971. Feeding Behavior in the Hawaiian Zoanthids *Palythoa* and *Zoanthus. Pacific Science* 25(4): 512–520. 3 figs.

Walsh, G. E.
    1967. *An Annotated Bibliography of the Families Zoanthidae, Epizoanthidae, and Parazoanthidae (Coelenterata, Zoantharia).* Univ. Hawaii—Hawaii Inst. Marine Biology Tech. Rep. 13. 77 pp.

Walsh, G. E., and R. L. Bowers
    1971. A Review of Hawaiian Zoanthids with Descriptions of Three New Species. *Zoological J. Linnean Soc.* 50(2): 161–180.

ORDER SCLERACTINIA
STONY CORALS
JAMES E. MARAGOS*

University of Hawaii

PRACTICALLY ALL of the 800 or more species of reef-building corals belong to the order Scleractinia (previously called Madreporaria). Although corals and reefs have been in existence since the Paleozoic, the scleractinian corals did not begin to evolve and assume dominance until late Mesozoic times.

GENERAL BIOLOGY

Corals, including scleractinians, are found in all the oceans of the world, and they differ from most other coelenterates in having a permanent skeleton. In sea fans, soft corals, and black corals, this skeleton is usually composed of proteinaceous substances. In scleractinian corals, some precious corals (including pink coral), and some hydrozoan coral, the skeleton is composed of calcium carbonate.

The thin living tissues of the colonial coral are made of a series of interconnected anemonelike polyps which cover the outer surfaces of the skeleton. Solitary corals have only a single polyp. Each polyp secretes a cuplike skeletal depression at its base called a calyx. The polyp contracts into the calyx when inactive or threatened.

Within the cells of the tissues of true reef-building corals (hermatypic corals) are found populations of single-celled algae called zooxanthellae, which are close relatives of the group of algae known as dinoflagellates. The zooxanthellae and coral host share a symbiotic relationship in which each member benefits the other. Like other algae, the zooxanthellae photosynthesize in daylight, producing organic matter from carbon dioxide and water. Some of the carbon dioxide originates from the respiration of the coral animal, but some may also be produced as a by-product during the formation of the coral skeleton. The removal of this carbon dioxide may increase the rate of carbonate deposition by the coral. This may be the reason hermatypic corals grow much more rapidly than ahermatypic corals.

---

*I would like to thank Dr. Keith Chave and Dr. Albert H. Banner for financial assistance, and Dr. Dennis Devaney for editorial assistance. Sue Monden, illustrator for the Department of Zoology, University of Hawaii, prepared the excellent line drawings. Dr. Arthur Reed provided the photographs for Figures 70, 78, 81, 83, 85–87, and 92 and Randi Schneider for Figure 60, and the latter also provided specimens of a few of rarer species. Dr. John W. Wells provided important information on the systematics of Hawaiian corals, and Richard H. Randall offered useful comments and criticisms on coral systematics. Other people who assisted in this report include Drs. Julie Bailey-Brock, Robert Johannes, John Randall, Lucius Eldredge, and Sidney Townsley. The Hawaii Institute of Marine Biology provided clerical assistance and photographic processing facilities.

The zooxanthellae in turn are thought to supply the coral with certain of their photosynthetic products, thus enabling the animal portion of the coral to be less dependent on food from outside sources. However, reef corals possess stinging tentacles, mouth parts, and a digestive system that are as well adapted as those of other coelenterates to catching and eating animal prey. However, we do not know whether the amount of food captured by the coral is sufficient to supply total food energy requirements, or whether the prey serves as a nutritional supplement, perhaps providing essential nutrients not available in the products of the zooxanthellae (Johannes and others, 1970).

The zooxanthellae gain other benefits from the relationship besides a continuous supply of carbon dioxide. By living within the coral tissues, the zooxanthellae are ensured a protected and stable environment. They may also gain inorganic nutrients from the coral which they can utilize to synthesize amino acids, proteins, and other important substances. The coral animal benefits from this, since the algae are functioning as "kidneys" for the organism. The internal recycling of nutrients also enables the reef coral to be less dependent upon extraneous sources of nutrients than are other autotrophs (food producing organisms). The relationship between the plant and animal components of the coral is so intricate that one may think of the association as a composite organism with a nature quite different from that of the two individual components.

The symbiotic relationship of hermatypic corals facilitates their rapid growth and dominance on reefs. With time, enormous carbonate reefs, composed of the skeletal remains of reef corals and other organisms may develop off coastlines or atop shoals in shallow water. Some branching reef corals may grow as fast as 10 cm per year, whereas massive forms may add 1 cm to 3 cm a year to their dimensions.

The growth of reef corals is in uneasy balance with a number of destructive forces. A host of fish and invertebrates browse or prey upon coral, including the crown-of-thorns starfish, *Acanthaster planci*. Other organisms including polychaete worms, sponges, mollusks, and filamentous algae, bore into and live within coral skeletons. Waves and storms can also damage coral. Dead coral skeletons in time may be overgrown by calcareous algae and cemented with other carbonate remains to form rigid reef rock or perhaps accumulate along shorelines where they are eroded by waves to form white sand beaches.

Where the reef is healthy, its upward growth is rapid enough to overcome destructive forces, as well as to keep up with any downward subsidence of underlying formations. Most hermatypic corals are confined to water depths of less than 100 m because zooxanthellae need adequate sunlight for photosynthesis. Reef corals are also restricted to the warm waters of tropical oceans. But reefs and reef corals may also be found outside the tropics, such as off Japan, where warm-water currents extend into temperate latitudes along the western boundaries of the world's major oceans.

Reef corals are usually found cemented to the bottom. However, a few reef corals, like the solitary mushroom corals *(Fungia)*, live unattached as adults. Waves and currents may transport unattached corals and broken living fragments of other corals from one reef site to another where they may act as seed populations for new coral communities.

Reef corals are more commonly dispersed between reefs, island groups, and even oceans during their free-swimming larval stage. The coral larva, or planula, is produced within the cavities of the adult polyps. Triggered by phases of the moon, tides, water temperatures, or other stimuli, the adults expel the larvae into the sea, where they drift as plankton for periods ranging from an hour to several months (see Harrigan, 1972). During this time ocean currents may carry the larvae to distant shores.

When the larvae come into contact with suitable, usually solid surfaces, they settle, attach, and begin life as adults. The single attached polyp soon begins to secrete a skeleton and, if of a colonial species, multiplies into a colony. On the average, only a few of the larvae expelled by polyps of each coral during its life will succeed in developing into mature adults.

Some reef corals can live for many years and grow to immense size. Indefinite growth may ultimately be limited by catastrophic events such as *Acanthaster* infestations, storm wave destruction, or sediment burial. Reef corals display remarkable powers of regeneration and repair. After damage, the polyps of a colony may act in unison to restore original colony shape within only a few years.

Reef corals can also live in a variety of forms on the reef with members of the same species often assuming different shapes, depending upon environmental conditions. This plasticity of growth form has created perplexing problems in coral identification.

In recent years man's activities and presence have increasingly threatened coral reefs, particularly in Hawaii. Especially damaging are sewage discharge, sedimentation, fresh-water floods, dredging activities, and fishing by dynamite. Other stresses include thermal, radiation, oil, and perhaps pesticide pollution. The crown-of-thorns starfish *Acanthaster* apparently is not a serious threat to Hawaiian reef corals (Branham and others, 1971).

Corals are particularly vulnerable to pollution since they cannot move out of polluted areas. Pollution destroys coral communities in ways that are often subtle, especially in environments where flushing and circulation are restricted. For example, sewage contains nutrients which stimulate the growth of suspended phytoplankton in the water column. The subsequent rain of organic material on the substrate requires large amounts of oxygen for its oxidation and chemical removal. The resultant oxygen depletion in the substratum may suffocate coral or cause it to be poisoned with hydrogen sulfide or other substances characteristic of environments lacking oxygen. Increased plant nutrients from sewage may also stimulate certain benthic algae to grow rapidly. Apparently reef corals are not similarly stimulated, since they probably need few extraneous nutrients to supplement their internal reserves, which are constantly recycled. Thus, benthic algae may eventually outcompete the reef corals and smother them. Recently, large populations of the bubble algae *(Dictyosphaeria)* have destroyed a large proportion of the finger coral communities in Kaneohe Bay, Hawaii. The explosion in the algae was coincident with the discharge of sewage into the bay (Banner and Bailey, 1970; Maragos, 1972).

Poor land management practices may also affect reef corals. Improper grading and planting of cleared hillsides results in increased runoff and soil erosion. This, in turn, increases the likelihood of flood damage and sedimentation on coral reefs near the mouths of drainage basins.

Coral reefs are a unique aesthetic, ecologic, and economic resource. Once coral communities are destroyed, recovery may take decades or more, even if the stresses are totally removed. Without corals, many of the fish and invertebrates that live on reefs would also disappear. As human populations continue to grow, it becomes increasingly important for man to develop laws and activities that will encourage preservation of reef resources, especially in Hawaii. A Hawaii state law prohibits taking coral from shallow water without a permit, except for domestic and noncommercial purposes.

REEF CORALS OF HAWAII

Hawaiian reef corals originated from the Indo-Pacific fauna, probably by way of Wake Island or the Line Islands, located to the southwest and south, respectively. The Hawaiian Islands lie near the periphery of the northern tropics where temperatures and light levels are lower than those of more tropical areas. Also the islands are isolated from the western Pacific where many of the reef corals originally evolved. As a consequence, the scleractinian coral fauna of Hawaii is represented by only 40 reef-building species belonging to 16 genera and subgenera. In comparison, more than 70 species occur at Fanning Atoll (1,600 km to the south) and more than 200 species at the Samoan and Marshall Islands farther to the southwest (Maragos, 1974; Wells, 1954; Hoffmeister, 1925; and others). Other evidence also indicates that reef corals grow more slowly in Hawaii than elsewhere (Maragos, 1972).

Not only is the Hawaiian coral fauna impoverished, but the present growth of the reefs themselves is poor or declining. Erosion is probably occurring on most Hawaiian reefs except those in bays or leeward coasts sheltered from wave action. Reef areas unaffected by pollution and other recent stresses also show signs of decline. Reefs must have been better developed in prehistoric times because, under present circumstances, the large thick reefs found both submerged and emerged around Kauai, Oahu, Molokai, Lanai, and many of the leeward atolls could not have been established. Other evidence further indicates that some fossil corals are no longer present in Hawaii (Menard and others, 1962). Hawaiian reefs lack steep ocean-facing slopes, a characteristic of many stable and actively growing reefs. Only a single small barrier reef is located in the islands (Kaneohe Bay Reef). Hawaiian corals rarely grow within 0.5 m of the sea surface, probably because of environmental extremes in temperature and other factors. In contrast, some reef corals are commonly exposed at nearly every low tide on more flourishing reefs outside Hawaii.

Vigorous growth of Hawaiian corals usually occurs between depths of 2 m to 10 m. The most flourishing coral communities occur on the leeward (western) sides of the islands where the erosive nature of wave action is reduced. Well-developed coral communities occurring on the windward sides are confined to protected bays. The poor development of Hawaiian reefs and corals may also be attributed to the normally high rainfall and runoff rates characterizing the windward sides of most high islands. The natural degradation of reef resources in Hawaii is now being augmented by pollution coincident with the recent increase in urbanization and population growth in the islands.

PREVIOUS RESEARCH ON THE SYSTEMATICS AND IDENTIFICATION OF HAWAIIAN SCLERACTINIA

More than 150 species of scleractinian corals, including synonyms, have been described or reported from the Hawaiian Islands. However, this report recognizes only 42 scleractinian species in shallow water (upper 100 m; see Table 1), and all but 2 species are reef corals.

TABLE 1
Adequately Described Scleractinian Corals with Documented Presence in Shallow Hawaiian Waters

Acroporidae
    Acropora paniculata Verrill, 1902
    Montipora dilatata Studer, 1901
    Montipora flabellata Studer, 1901
    Montipora patula Verrill, 1864
    Montipora verrilli Vaughan, 1907
    Montipora verrucosa (Lamarck, 1816)

Agariciidae
  *++*Leptoseris hawaiiensis,* Vaughan, 1907
    Leptoseris incrustans (Quelch, 1886)
    Leptoseris papyracea (Dana, 1846)
   **Leptoseris scabra* Vaughan, 1907
   **Leptoseris tubulifera* Vaughan, 1907
    Pavona duerdeni Vaughan, 1907
    Pavona varians Verrill, 1864
    Pavona (Pseudocolumnastraea) pollicata Wells, 1954

Balanophyllidae
  *+*Balanophyllia* spp. cf. *B. affinis* (Semper, 1872) and *B. hawaiiensis* Vaughan, 1907

Dendrophyllidae
   +*Tubastraea coccinea* Lesson, 1831

Faviidae
    Cyphastrea ocellina (Dana, 1846)
    Leptastrea bottae (Milne-Edwards and Haime, 1850)
    Leptastrea purpurea Dana, 1846

Fungiidae
    Cycloseris fragilis (Alcock, 1902)
    Cycloseris hexagonalis (Milne-Edwards and Haime, 1849)
    Cycloseris vaughani (Boschma, 1923)
    Fungia (Pleuractis) scutaria Lamarck, 1801

Pocilloporidae
    Pocillopora damicornis (Linnaeus, 1758)
    Pocillopora eydouxi Milne-Edwards and Haime, 1860
    Pocillopora ligulata Dana, 1846
    Pocillopora meandrina Dana var. *nobilis* Verrill, 1864
    Pocillopora molokensis Vaughan, 1907

Poritidae
    Porites brighami Vaughan, 1907
    Porites compressa Dana, 1846
    Porites duerdeni Vaughan, 1907
    Porites evermanni Vaughan, 1907
    Porites lichen Dana, 1846
    Porites lobata Dana, 1846
    Porites pukoensis Vaughan, 1907
   **Porites studeri* Vaughan, 1907
    Porites (Synaraea) convexa Verrill, 1864
    Porites (Synaraea) irregularis Verrill, 1864

Siderastreidae
    Coscinaraea ostreaeformis Van der Horst, 1922

Thamastreidae
    Psammocora nierstraszi Van der Horst, 1922
    Psammocora verrilli Vaughan, 1907
    Psammocora (Stephanaria) stellata Verrill, 1864

---

*Confined to deep reefs.
+Ahermatypic (non-reef-building) corals.
++Ahermatypic for most of its depth range, but may contain zooxanthellae algae at depths above 150 m.

TABLE 2
Previously Reported Reef Corals from Hawaii of Doubtful Existence*

Acropora echinata Dana [2, 4]
Alveopora dedalea (Forskal) [1, 2] = *Alveopora verrilliana* Dana
Alveopora verrilliana Dana [2]
Astraea (Orbicella) ocellina Dana [1] = *Cyphastrea ocellina* (Dana)
Astraea rudis Studer [1, 2] = *Favia speciosa* (Dana)
Coelastrea tenuis Verrill [1] = *Leptastrea purpurea* (Dana)
Cycloseris patelliformis (Boschma) [2]
Dendrophyllia manni (Verrill) [1] = *Tubastraea coccinea* Lesson
Favia hawaiiensis Vaughan [1] = *Leptastrea purpurea* (Dana)
Favia hombroni (Rousseau)? [1, 2] = *Favia stelligera* (Dana) [2]
Favia rudis Verrill [1, 2] = *Favia speciosa* (Dana)
Favia speciosa (Dana) [2]
Fungia dentigera Leuckart [1] = *Fungia (Pleuractis) scutaria* Lamarck
Fungia echinata (Pallas) [2]
Fungia fragilis Alcock [1] = *Cycloseris fragilis* (Alcock)
Fungia oahensis Doderlein [1] = *Fungia (Pleuractis) scutaria* Lamarck
Fungia patella (Ellis and Solander) [1] = *Cycloseris vaughani* (Boschma)
Fungia paumotensis Stutchbury [2]
Leptastrea agassizi Vaughan [1] = *Leptastrea bottae* (Milne-Edwards and Haime)
Leptastrea hawaiiensis Vaughan [1] = *Leptastrea bottae* (Milne-Edwards and Haime)
Leptastrea stellulata Verrill [1] = *Leptastrea purpurea* (Dana)
Leptoseris digitata Vaughan [1] = *Leptoseris papyracea* (Dana)
Lobactis danae Verrill [1] = *Fungia (Pleuractis) scutaria* Lamarck
Montipora bernardi Vaughan [1] = *Montipora verrucosa* (Lamarck)
Montipora capitata Dana [1] = *Montipora verrucosa* (Lamarck)
Montipora incognita Bernard [1] = *Montipora patula* Verrill
Montipora studeri Vaughan [1] = *Montipora verrucosa* (Lamarck)
Montipora tenuicaulis Vaughan [1] = *Montipora verrucosa* (Lamarck)
Montipora venosa (Ehrenberg) [5] (probably *Montipora flabellata* Studer of Dana (1971)
Mussa? sp. young? Vaughan [5] (*Madrepora kauaiensis* Vaughan)
Pavona clavus (Dana) [5] (*Pavona duerdeni* Vaughan of Dana 1971)
Pavona explanulata Lamarck [1, 3] = *Pavona duerdeni* Vaughan
Pavona repens Brueggemann? [1] = *Pavona varians* Verrill
Pocillopora aspera Verrill [1] = *Pocillopora ligulata* Dana
Pocillopora brevicornis Verrill [1] = *Pocillopora damicornis* (Linnaeus)
Pocillopora cespitosa Dana [1] = *Pocillopora damicornis* (Linnaeus)
Pocillopora elegans Dana [5] (*Pocillopora meandrina* Dana of Franzisket 1970)
Pocillopora elongata Dana [1] = *Pocillopora eydouxi* Milne-Edwards and Haime
Pocillopora favosa Ehrenberg [1] = *Pocillopora ligulata* Dana
Pocillopora frondosa Verrill [1] = *Pocillopora ligulata* Dana
Pocillopora informis Dana [6] (probably wave stressed form of *Pocillopora meandrina* or *Pocillopora ligulata* Dana)
Pocillopora modumanensis Vaughan [1] = *Pocillopora eydouxi* Milne-Edwards and Haime
Pocillopora nobilis Verrill [1] = *Pocillopora meandrina* Dana
Pocillopora plicata Dana [1] = *Pocillopora ligulata* Dana

(Cont. next page)

---

*Numbers indicate reasons for doubting the existence of the species in Hawaii:
1. Synonym of another species (followed by preferred name).
2. Unauthenticated or doubtful records for the Hawaiian Islands, or specimens lost.
3. Poorly described species.
4. Extremely rare, if at all present in Hawaii.
5. Incorrect identification (correct name and relevant reference in parentheses).
6. Form of described species severely modified by environmental stress and probably belongs to another species (in parentheses).
7. Colony size of type specimen too small or young to merit separate rank from related species (in parentheses).

TABLE 2 (Cont.)*

Pocillopora rugosa Gardiner[1] = *Pocillopora eydouxi* Milne-Edwards and Haime
Pocillopora solida Quelch[1] = *Pocillopora molokensis* Vaughan
Pocillopora verrucosa Dana[1] = *Pocillopora meandrina* Dana
Porites bulbosa Quelch[1] = *Porites compressa* Dana
Porites compressus[5] (*Porites compressa* Dana of Hobson, 1974)
Porites lanuginosa Studer[3]
Porites lichen Quelch[1] = *Porites lobata* Dana
Porites mordax Dana[1,3] = *Porites compressa* Dana
Porites pukoensis[5] (*Porites lobata* Dana of Hobson, 1974)
Porites quelchi Studer[1] = *Porites lobata* Dana
Porites reticulosa Dana[1,2] = *Porites lichen* Dana
Porites schauinslandi Studer[2,3]
Porites tenuis Quelch[1,2] *Porites lobata* Dana
Porites (Synaraea) hawaiiensis Vaughan[7] (*Porites (Synaraea) convexa* Verrill)
Psammocora (Stephanaria) brighami Vaughan[1] = *Psammocora (Stephanaria) stellata* Verrill
Tubastrea aurea Quoy and Gaimard[1] = *Tubastraea coccinea* Lesson

*See footnotes previous page.

TABLE 3
AHERMATYPIC (NON-REEF-BUILDING) SCLERACTINIAN CORALS
CONFINED TO DEEP WATER IN HAWAII

Anisopsammia amphelioides (Alcock)—40*
Anthemiphyllia pacifica Vaughan—92
Balanophyllia desmophyllioides Vaughan—78
Balanophyllia diomedeae Vaughan—148
Balanophyllia hawaiiensis Vaughan—190
Balanophyllia laysanensis Vaughan—130
Bathyactis hawaiiensis Vaughan—963
Caryophyllia alcocki Vaughan—876
Caryophyllia hawaiiensis Vaughan—92
Caryophyllia octopali Vaughan—281
Ceratotrochus laxus Vaughan—319
Cyathoceras diomedeae Vaughan—169
Deltocyathus andamanicus Alcock—147
Dendrophyllia oahensis Vaughan—154
Dendrophyllia serpentina Vaughan—147
Desmophyllum cristagalli Milne-Edwards and Haime—220
Endopachys oahense Vaughan—53
Flabellum deludens v. Marenzeller—670
Flabellum pavoninum Lesson—127
Gardineria hawaiiensis Vaughan—272
Madracis kauaiensis Vaughan—24
Madrepora kauaiensis Vaughan—294
Paracyathus gardineri Vaughan
Paracyathus mauiensis Vaughan—95
Paracyathus molokensis Vaughan—88
Paracyathus tenuicalyx Vaughan—252
Placotrochus fuscus Vaughan—148
Stephanophyllia formosissima Moseley—66
Trochocyathus oahensis Vaughan—252

*Numbers refer to the shallowest reported depths (in fathoms). Data summarized from T.W. Vaughan (1907). Names of species also follow those of T.W. Vaughan (1907).

Before 1907 a combined total of 38 species of scleractinians, including synonyms, was recorded from Hawaii in the studies by Lesson (1831), J. D. Dana (1846), Verrill (1864, 1865), Quelch (1886), Fowler (1888), Brook (1893), Bernard (1897, 1905), Studer (1901), and T. W. Vaughan (1905). T. W. Vaughan's (1907) definitive treatise on Hawaiian corals included descriptions of a number of new species and records as well as a thorough revision and analysis of previous work. The report increased the total Hawaiian coral fauna to 86 species and 129 varieties and forms. Subsequent studies (Boschma, 1954; Wells, 1954) have considered about 20 of these species to be synonyms of other preferred names, unauthenticated records, poorly described species, or extremely rare (Table 2). Of the remainder, at least 33 are considered to be ahermatypic corals (Table 3) with only two or three present on shallow reefs. Hence, 33 of the species reported by T. W. Vaughan (1907) are recognized here as reef corals with authenticated distribution in the shallow waters of Hawaii. Since Vaughan's treatise, 7 additional species have been recorded (T. F. Dana, 1971; Reed, 1971; Maragos, 1972, this report). The Hawaiian reef-building coral fauna is now represented by 40 species and 16 genera and subgenera.

Other important systematic studies on Hawaiian corals include Powers' (1970) numerical taxonomic analysis of 20 common species and Powers and Rohlf's (1972) systematic study of Hawaiian and Caribbean corals.

In the following key to Hawaiian corals, the reader should make use of the glossary of technical terms, the figures indicated, as well as the descriptions of species which follow. Color characters refer to living corals.

If it is necessary to examine the morphological features of the coral skeleton in detail or to preserve the coral in a dry condition, the soft tissue should be removed. This can be done by soaking the coral in equal parts of 5 to 6 percent sodium hypochlorite (common bleach) and fresh water, followed by a fresh-water rinse (agitation is sometimes necessary to remove deeply embedded material), and then drying prior to examination.

KEY TO CORALS KNOWN FROM HAWAIIAN WATERS

1   Colonial coral; skeleton has many calices 7
       Solitary coral; skeleton has only a single calyx or a portion of one 2

2(1)   Coral attached to substratum, frequently by stalks 3
       Coral unattached to the substratum, skeleton a disk or a radial 4

3(2)   The head or upper end of the coral is disklike, between 2 cm and 4 cm in diameter; found in shallow water especially on some reef flats; common. These represent the immature stage of the solitary coral *Fungia (Pleuractis) scutaria* Lamarck.
       Upper end of coral is not a disk less than 2 cm in diameter; coral resembles a hollow cylinder with 4 cycles of septa of nearly equal dimension; deep water, rare. .... *Balanophyllia* sp [Figs. 1, 69] (probably *B. hawaiiensis* Vaughan and/or *B. affinis* (Semper))

4(2)   Coral has circular outline; septa thickened at bases; upper surface convex, lower surface concave; mouth is a deep slit centrally located; moderate to deep water; pale brown or pale green. ....
       .... *Cycloseris vaughani* (Boschma) [Figs. 2, 82, 83]

SCLERACTINIA

Figure 1.—*Balanophyllia* sp. Compare *B. hawaiiensis* Vaughan.

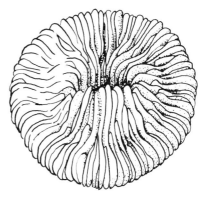

Figure 2.—*Cycloseris vaughani* (Boschma).

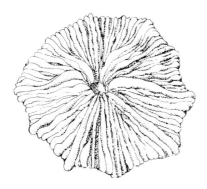

Figure 3.—*Cycloseris hexagonalis* Milne-Edwards and Haime.

Figure 4.—*Fungia (Pleuractis) scutaria* Lamarck.

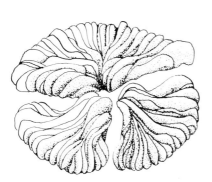

Figure 5.—*Cycloseris fragilis* (Alcock).

Figure 6.—*Pavona (Pseudocolumnastraea) pollicata* Wells.

         Skeleton elliptical, polygonal, indented, deeply incised to one side, or
            a small portion of a disk; septa thin and numerous. 5
5(1)   Disk outline is polygonal or angular, upper and lower surfaces are flat;
            primary septa higher than rest; maximum diameter less than 2 cm;
            rare.............. *Cycloseris hexagonalis* Milne-Edwards and Haime
                                                                    [Figs. 3, 86, 87]
         Skeleton rounded or a radial segment (like a piece of pie) 6
6(5)   Disk outline elliptical, indentations absent; coral large and greater than
            4 cm in diameter; upper surface convex, lower is concave; mouth is
            long and 1/5 the total length; many razor-thin septa; light to dark
            brown, sometimes with green tentacles; common. ....
            .... *Fungia (Pleuractis) scutaria* Lamarck [Figs. 4, 78–81]
         Disk margin indented, deeply incised, or a radial segment, like a piece of
            pie; less than 4 cm in diameter; upper and lower surfaces flat; septa
            beaded and not continuous; septa of variable height; rare............
            .... *Cycloseris fragilis* (Alcock) [Figs. 5, 84, 85]
7(1)   Septa do not continue outside calyx boundaries as septo-costae; calices
            distinct from one another.................................... 19
         Septa continue out beyond the boundary of the calyx into the coenosteum
            as septo-costae; adjacent calices are not separated by walls
            or ridges. 8
8(7)   Circular calices elevated; large rounded cavities present in the center of
            the calices; septo-costae well developed, especially on calyx walls;
            colonies foliaceous or branching. ....
            .... *Pavona (Pseudocolumnastraea) pollicata* Wells [Figs. 6, 59, 60]
         Calices in depressions; center of calices have small cavities. 9
9(8)   Septo-costae beaded or granulated; colony form encrusting or ramose
            clumps... 16
         Septo-costae compact or laminar. 10
10(9)   Septo-costae are as thin as hair, frequently show curves or bends, and
            are longer than 1 to 2 cm. 11
         Septo-costae thicker than hair, straight, and not longer than 1 cm. 15
11(10) Colony margin not free; colony form encrusting, showing irregular
            knobby surface; calices in depressions between swellings; moderate
            to deep water .... *Leptoseris incrustans* (Quelch) [Figs. 7, 62]
         Colony margin free and very thin; colony form explanate, flowerlike,
            or cuplike. 12
12(11) Portions of the margin bend around to meet and fuse into tubes which
            grow upright; deep water; rare. .... *Leptoseris tubulifera* Vaughan
                                                                     [Figs. 8, 67, 68]
         Margin straight or undulating, not closing off to form tubes. 13
13(12) Margin wavy and forming lobes and sublobes; colony looks like a flower;
            some calices in lobes; deep water........ *Leptoseris papyracea* (Dana)
                                                                     [Figs. 9, 63, 64]
         Margin does not form regular lobes; colony surface irregular or
            cuplike. 14

SCLERACTINIA

Figure 7.—*Leptoseris incrustans* (Quelch).

Figure 8.—*Leptoseris tubulifera* Vaughan.

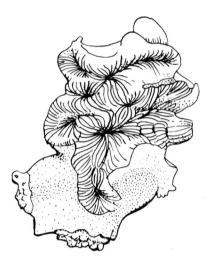

Figure 9.—*Leptoseris papyracea* (Dana).

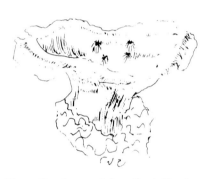

Figure 10.—*Leptoseris hawaiiensis* Vaughan.

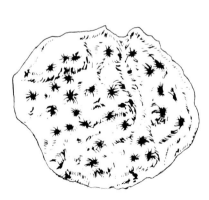

Figure 11.—*Leptoseris scabra* Vaughan.

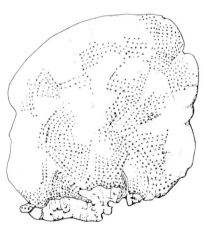

Figure 12.—*Pavona duerdeni* Vaughan.

14(13) Upper surface flat; margin sometimes turned upward so that colony resembles a broad shallow cup; calices in depressions; very deep water. .... *Leptoseris hawaiiensis* Vaughan [Figs. 10, 65]
Upper surface covered with gentle hills; calical depressions located at the crests of hills; deep water. ....
.... *Leptoseris scabra* Vaughan [Figs. 11, 66]

15(10) Colony surface smooth; colony form encrusting or glomerate or vertically flattened lobes; calices arranged in regular symmetrical pattern; light gray or pale gray green ....
.... *Pavona duerdeni* Vaughan [Figs. 12, 57, 58]
Surface of coral covered with elevated ridges, some which are elongated and have steep walls; top edge of ridges not rounded; calices nestled in V-shaped valleys between ridges; colony form encrusting or foliaceous; tan or brown. ....
.... *Pavona varians* Verrill [Figs. 13, 54–56]

16(9)   Calices sit in slight mounds; colony form encrusting or foliaceous; septo-costae thick, long, and uniformly beaded, beads the size of coarse sand; white to tan...
.... *Coscinaraea ostreaeformis* Van der Horst [Figs. 14, 113]
Calices flush with surface, in shallow funnel-like depressions, or in valleys between ridges, septo-costae finely beaded or granulated. ... 17

17(16) Colony form ramose, branching clumps or irregular encrustations; each calyx in a shallow funnel-like crater. ....
... *Psammocora (Stephanaria) stellata* Verrill [Figs. 15a, 15b, 114–116]
Colony form encrusting; calices in depressions; ridges sometimes elongated or meandering...................................... 18

18(17) Surface of coral not covered with toothlike projections the size of sand; ridges are slight and rounded; individual calices in separate depressions; gray or pale brown; rare. ....
.... *Psammocora verrilli* Vaughan [Figs. 16, 117]
Toothlike projections on all surfaces, giving a rough sandpaper texture; ridges are elongate meandroid or crescentic and have steep sides; green or dark brown. ....
.... *Psammocora nierstraszi* Van der Horst [Figs. 17, 118]

19(7)   Calices separated from one another by a coenosteum space. 20
Calices polygonal, crowded together, and connected by networks of common walls, ridges, seams, or grooves. 32

20(19) Calices lack elevated walls, funnel-like or resemble stars or snowflakes.. 21
Calices round and have walls elevated above the coenosteum (or general colony surface). 29

21(20) Minute rodlike tubercles or nipplelike papillae cover surface; projections are irregular, rounded, or crescent-shaped lobes if present. 22
Colony surface lacks projections and appears smooth. 26

22(21) Ridges irregular, sometimes elongated, crescentic, and rounded, calices aggregated in depressions between ridges; colony form foliaceous

SCLERACTINIA

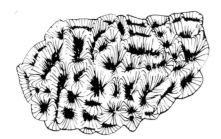

Figure 13.—*Pavona varians* Verrill.

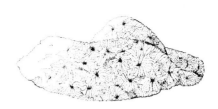

Figure 14.—*Coscinaraea ostreaeformis* Van der Horst.

Figure 15.—a, *Psammocora (Stephanaria) stellata* Verrill.

Figure 15.—b, *Psammocora (Stephanaria) stellata* Verrill.

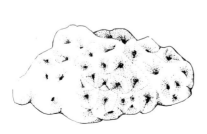

Figure 16.—*Psammocora verrilli* Vaughan.

Figure 17.—*Psammocora nierstraszi* Van der Horst.

	or columniform; charcoal brown. ....
	.... *Porites (Synaraea) convexa* Verrill [Figs. 18, 110, 111]
	Projections nipplelike papillae or minute rodlike tubercles. 23
23(22)	Projections tubercles the size of pinheads; tan, sometimes polyps dark blue. 25
	Projections papillae 0.3 cm to 0.5 cm in diameter; brown or light blue.... 24
24(23)	Papillae absent from undersurfaces of free margins, if present; each calyx a small, deep pit situated next to a papilla; colony form extremely variable; beige to chocolate brown. ....
	.... *Montipora verrucosa* (Lamarck) [Figs. 19a, 19b, 46, 47]
	Papillae small (0.2 cm to 0.3 cm) and sometimes fuse to form crescentic ridges; calices situated in grooves or crevices have an elevated collar; colonies encrusting and lack free margin; bright blue or sometimes brown or red. .... *Montipora flabellata* Studer [Figs. 20, 49]
25(23)	Minute rodlike tubercles all of same height rarely aggregate together; star-shaped calices depressed at bases of many tubercles; colonies encrusting and lack free edges............................
	.... *Montipora verrilli* Vaughan [Figs. 21, 52]
	Every calyx surrounded by a ring of small fused tubercles; calices small; tubercles with same diameters but different heights; some calices depressed, others elevated; colony form encrusting or foliaceous; surface appears rough or irregular; tan...........................
	.... *Montipora patula* Verrill [Figs. 22, 50, 51]
26(21)	Calices small, star-shaped, deep cavities; colony form foliaceous or glomerate; each calyx with a ringlike collar; chocolate brown; rare..................... *Montipora dilatata* Studer [Figs. 23, 48]
	Calices round, small, and resemble frosted snowflakes. 27
27(26)	Colony form encrusting; calices very small and tightly packed; colonies less than 10 cm; shallow water, especially reef flats.................
	.... *Porites lichen* Dana
	Colony form lobose, collumniform, or platelike; calices crowded to widely spaced. 28
28(27)	Colony form resembles a cauliflower; branches fused at tips to give an irregular nodular appearance; rare. ....
	.... *Porites (Synaraea) irregularis* Verrill [Figs. 24, 112]
	Colony form foliaceous or sometimes columniform; charcoal brown; rare.......... *Porites (Synaraea) convexa* Verrill [Figs. 25, 110, 111]
29(20)	Calices large (1 cm diameter) and have elevated walls which form tubes 2 cm long or more; orange, vermillion, or sometimes black. ....
	.... *Tubastraea coccinea* Lesson [Figs. 26, 70, 71]
	Calices less than 1 cm in diameter, walls less than 0.5 cm high. 30
30(29)	Calices all small and less than 0.5 cm diameter; calices have high, jagged walls with striations, calices crowded and point randomly in many directions; colony form encrusting or glomerate; colonies small (less than 15 cm); reddish brown. ....
	.... *Cyphastrea ocellina* (Dana) [Figs. 27, 72]

SCLERACTINIA

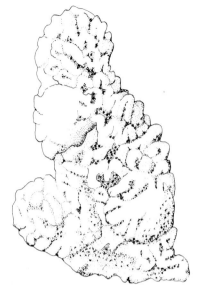

Figure 18.—*Porites (Synaraea) convexa* Verrill.

Figure 19.—*a, Montipora verrucosa* (Lamarck).

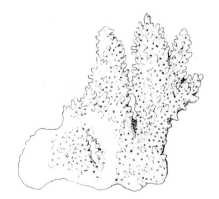

Figure 19.—*b, Montipora verrucosa* (Lamarck).

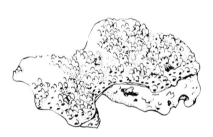

Figure 20.—*Montipora flabellata* Studer.

Figure 21.—*Montipora verrilli* Vaughan.

Figure 22.—*Montipora patula* Verrill.

SCLERACTINIA

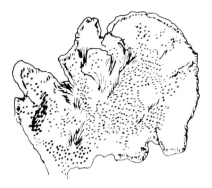

Figure 23.—*Montipora dilatata* Studer.

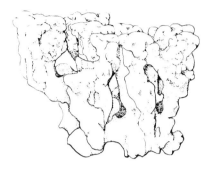

Figure 24.—*Porites (Synaraea) irregularis* Verrill.

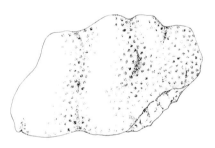

Figure 25.—*Porites (Synaraea) convexa* Verrill; small, immature colony.

Figure 26.—*Tubastraea coccinea* Lesson.

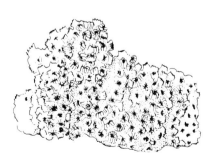

Figure 27.—*Cyphastrea ocellina* (Dana).

Figure 28.—*Leptastrea bottae* (Milne-Edwards and Haime).

	Calices generally large (greater than 0.5 cm); walls are thin and short; primary septa thick and sometimes elevated above surface of calyx wall; olive brown. . . . . . *Leptastrea bottae* (Milne-Edwards and Haime) [Figs. 28, 73, 74]
31(19)	Calices large (up to 1 cm diameter), polygonal, crowded, depressed, and funnel-like, adjacent calices have common walls sometimes separated by a fine groove; colony form encrusting; light brown, green or purple. . . . . *Leptastrea purpurea* Dana [Figs. 29, 75–77]
Calices small, less than 0.5 cm in diameter. 32
32(31)	Calices small (1 mm), angular, resemble pinholes; calices simple deep holes lacking development of septa, walls of calices jagged, thin, and flare out; calices crowded together on wartlike projections [Fig. 30] (verrucae) of 0.5 cm diameter; in other colonies, the verrucae are replaced by long, slender branches of same thickness; all forms ramose or cespitose (branching). 40
Calices larger (2 mm to 4 mm), five-sided polygons; intricate development of septa and other structures give calices a frosted snowflake appearance [Fig. 31] 33
33(32)	Colonies encrusting, glomerate, or explanate. 35
Colonies show branching or elongated meandering folds. 34
34(33)	Branches are thick, irregular, show numerous constrictions and swellings; calices small (less than 2 m diameter), crowded; rare. . . . .
. . . . *Porites duerdeni* Vaughan [Figs. 32, 101]
Some branches fingerlike and cylindrical, others fused into meandering folds 2 cm thick; length of branches variable and often branches are fused; very common. . . *Porites compressa* Dana [Figs. 33, 99, 100]
35(33)	Calices depressed between narrow ridges or are funnel-like 37
Calices very shallow or flush with surface 36
36(35)	Calices appear to be surrounded by a double concentric ridge and are flush with surface; growth form cuboidal with flat topped lobes and rounded edges; colonies large, confined to shallow water
. . . . *Porites evermanni* Vaughan [Figs. 34, 104, 105]
Calices shallow funnels having single ridges or walls; growth form subspheroid (ball-like); colonies small and confined to deep water . . . . *Porites studeri* Vaughan [Figs. 35, 106]
37(35)	Calices large (3 mm to 4 mm), deep, funnel-like; septa not extending far toward the center of calyx; colony form small, encrusting; common in shallow wave exposed environments
. . . . *Porites brighami* Vaughan [Figs. 36, 108]
Calices small to medium (less than 3 mm), show intricate development of septa and other structures 38
38(37)	Calices separated by a thin ridge network; colonies encrusting, glomerate, or hemispherical; sometimes colonies show low cones or erect columns, never branching . . . . *Porites lobata* Dana [Figs. 37, 102, 103]
Colonies show tall, irregular, fingerlike lobes or composed of tall, pointed cones fused at bases 39

SCLERACTINIA

Figure 29.—*Leptastrea purpurea* Dana.

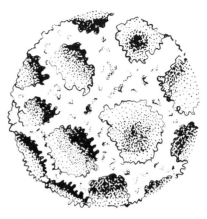

Figure 30.—Close-up of *Pocillopora*.

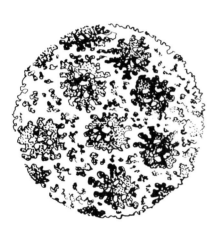

Figure 31.—Close-up of *Porites*.

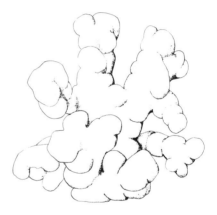

Figure 32.—*Porites duerdeni* Vaughan.

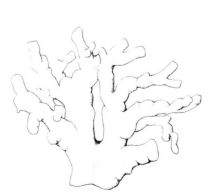

Figure 33.—*Porites compressa* Dana.

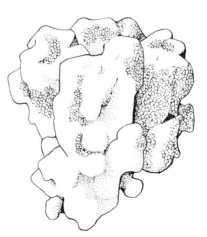

Figure 34.—*Porites evermanni* Vaughan.

SCLERACTINIA

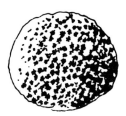

Figure 35.—*Porites studeri* Vaughan.

Figure 36.—*Porites brighami* Vaughan.

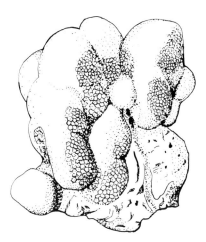

Figure 37.—*Porites lobata* Dana.

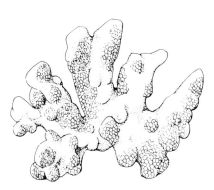

Figure 38.—*Porites compressa* Dana.

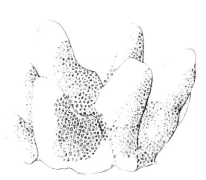

Figure 39.—*Porites pukoensis* Vaughan.

Figure 40.—*a, Pocillopora damicornis* (Linnaeus).

SCLERACTINIA

Figure 40.—b, *Pocillopora damicornis* (Linnaeus).

Figure 41.—*Pocillopora eydouxi* Milne-Edwards and Haime.

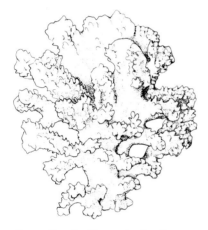

Figure 42.—*Pocillopora ligulata* Dana.

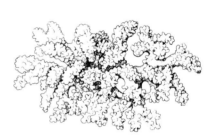

Figure 43.—*Pocillopora molokensis* Vaughan.

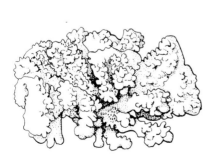

Figure 44.—a, *Pocillopora meandrina* Dana.

Figure 44.—b, *Pocillopora meandrina* Dana.

39(38) Lobes irregular, sometimes branchlike, often fused along their sides
.... *Porites compressa* Dana [Figs. 38, 99, 100]
Tall, conical lobes with pointed crests, 5 cm to 10 cm high; calices deep and large; not common .... *Porites pukoensis* Vaughan [Figs. 39, 107]

40(32) Branches with many divisions, no thicker than a pencil; colony form small, cespitose clumps (less than 15 cm); restricted to shallow water (less than 10 m); common
.... *Pocillopora damicornis* (Linnaeus) [Figs. 40a, 40b, 89–92]
Branches with few to several divisions, thicker than a pencil; colony form ramose; many calices are crowded on wartlike projections (verrucae) 41

41(40) Branches are long, cylindrical, and thick (3 cm); colonies sometimes tall (60 cm), resemble moose antlers; some calices possess a columella; brown; moderate to deep water
..... *Pocillopora eydouxi* Milne-Edwards and Haime [Figs. 41, 97, 98]
Branches variable, never cylindrical; colonies never taller than 40 cm; pale brown, pink, or green 42

42(41) Some calices possess a columella; branches thin, wide, and join together at bases to form long, wavy folds; wartlike verrucae not prominent, their density decreases toward the base of coral; less than 30 cm diameter; shallow water ...... *Pocillopora ligulata* Dana [Figs. 42, 95]
All calices lack columella and are simple deep pits; verrucae prominent; branches massive, thicker than 1.5 cm in any dimension........... 43

43(42) Branches ribbonlike, growing more horizontally than vertically, branches have several divisions; colony resembles a flat bush, 30 cm to 70 cm in diameter; moderate to deep water; rare
.... *Pocillopora molokensis* Vaughan [Figs. 43, 96]
Branches flattened vertically, grow upright; branches thick, show few divisions, resemble thick leaves; tops of branches smooth, truncated; space between branches meandroid; width at branch ends sometimes greater than at bases; very common; shallow to deep water
Pocillopora meandrina Dana var. *nobilis* Verrill [Figs. 44a, 44b, 93, 94]

Family **Acroporidae**

Acropora. On most tropical islands members of *Acropora* are extremely abundant and assume a bewildering array of forms. In Hawaii, however, the genus is extremely rare. The first reports of *Acropora* from Hawaii were reviewed by Quelch (1886) and determined to be incorrect. Additional reports of living *Acropora* in the Hawaiian Islands were made by Brock (1893) and Studer (1901), based upon specimens determined as *A. echinata* (Dana). Studer's specimen was reviewed and figured by T. W. Vaughan (1907). *A. echinata* is known from Fiji, the Philippines, Samoa, and the Marshall Islands (Wells, 1954). No recent records of this species from Hawaii are known and Dr. John W. Wells of Cornell University is of the opinion that all these previous records of *Acropora* are incorrect, a view also shared by Verrill (1902).

Edmondson (1946b, p. 41, Fig. 20) did, however, record and figure an undetermined tabular species of *Acropora* from "comparatively shallow water" collected at French Frigate Shoals in the leeward (northwestern) Hawaiian Islands in 1936.

 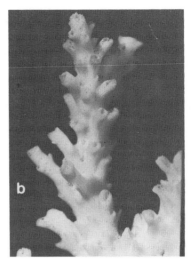

Figure 45.—SCLERACTINIA. A specimen of *Acropora* aff. *paniculata* Verrill collected at Poipu, Kauai, 1975: *a*, entire colony; *b*, terminal branch of same (courtesy of Dr. John W. Wells) (all photographs are of dead skeletal figures except where otherwise noted).

Unfortunately the preserved specimen has not been located, and specific identification cannot be made.

Quite recently, in 1975, a live *Acropora* colony was collected at a depth of 13 m off Poipu Beach, Kauai (Figs. 45*a*, 45*b*). Examination of the specimen by Dr. Wells suggests that it is a form closely related to *A. paniculata* Verrill or perhaps a new species. However, available coral material is insufficient to make a positive identification or describe it as a new species. Dr. Wells also believes that the growth form of the Poipu *Acropora* specimen is the same as that of the unidentified *Acropora* from French Frigate Shoals figured in Edmondson (1946b). If the Poipu *Acropora* is a new species, then it possibly represents part of a relict population of the species in Hawaii which is extremely rare, perhaps endemic, and possibly declining from some pre-existing eminence.

The Kauai specimen was collected by two University of Hawaii undergraduate students. The author examined the site and determined it to be characterized by large rounded basalt boulders, small sand patches, and a thin veneer of reef rock over solid basalt. The habitat was not particularly favorable to coral development because the coral cover was low and dominated by common wave-resistant species of *Porites* and *Pocillopora*.

In April, 1977, a colony of *Acropora* was reported at a depth of 13 m off the northwest side of La Perouse Pinnacle, French Frigate Shoals. A small portion of the colony was collected, and its skeletal features closely resemble those of the Poipu *Acropora* specimen, suggesting that both belong to the same species. The collector (John Naughton, pers. comm.) indicated the whole colony of the La Perouse *Acropora* measured 50 cm across, was located in a relatively barren coral area, and resembled the form of the *Acropora* figured in Edmondson (1946b).

It is of interest to note that *Acropora* exists in abundance at Johnston Island, only some 1,200 km to the southwest of Hawaii, and at the Line Islands, some 1,600 km to the south of Hawaii. Fossil skeletons of *Acropora* have also been reported on several occasions on Oahu. Menard and others (1962) reported both fossil *Acropora* and the brain coral *Platygyra* (the latter has not been reported live from Hawaii) on a submarine plateau south of Waikiki. Ken Roy and the author have noted fossil *Acropora* on uplifted reefs at Nanakuli and Kahuku, respectively.

Two major theories, or a combination thereof, can be put forth to explain the near absence of *Acropora* in Hawaii. One proposes that present climatic conditions are not favorable for the development of *Acropora* and many other common Indo-Pacific genera. Hawaii lies near the subtropics and light and temperature conditions may be suboptimal, at least during winter months. The other theory proposes that the islands are geographically isolated from the diverse coral areas to the southwest. As a consequence, the larvae of some coral genera never reach the islands because of the long distances between Hawaii and the nearest island groups. The colonization by some genera may occur so infrequently that the populations have not had the opportunity to become well established and inevitably die out with time.

The reported Hawaiian specimens of *Acropora echinata* and *A. paniculata* have a finely branched growth form with many small raised calices. Each calyx appears as a tube which projects above the surrounding skeletal space. The outer edge of the tube is rounded and resembles a trough or gutter because of the uneven projection of the tube walls. The minute tubelike calices of *Acropora* easily distinguish the genus from the others of the family. The skeleton is usually very porous and fragile. Although color information from the Hawaiian specimens is lacking, *Acropora* elsewhere usually assumes a beige or light brown color when alive.

Montipora. The genus *Montipora* is well represented in Hawaii, and at least 5 species and several varieties are commonly reported on reefs. Some species have variable growth forms, especially *Montipora verrucosa*, but all species are encrusting to some extent and a few assume a foliaceous form. Characteristic of the genus *Montipora* are numerous, small, deep, and star-shaped calices which usually lie in depressions between nipplelike or rodlike projections.

Montipora verrucosa (Lamarck) [syns. *M. bernardi* Vaughan, *M. studeri* Vaughan, *M. tenuicaulis* Vaughan]. This species is very common on the reefs in all environments between depths of 0 m to 50 m. The coral assumes a bewildering number of forms—microatoll, glomerate, pinnacle, branching, encrusting, or bracketlike plates (Fig. 46). One unusual form which may later be determined to be a separate variety shows deeply excavated (foveolate) calices and fingerlike branches. The species is easily identified by the numbers of nipplelike papillae covering most upper surfaces. Usually a calyx is found adjacent to each papilla (Fig. 47). The calices are open holes about 1 mm in diameter. The sides of the papillae descend directly into the depths of the cups, which are deep and have straight walls. Rodlike septa protrude only a short distance toward the center of each calyx, giving it a starlike appearance. The unusual variation in growth form and other characters had led to the description of at least four other species which were later determined to be synonyms of *M. verrucosa* (Table 2; and Boschma, 1954). Color varies from dark chocolate brown with white borders on the margins and on the crests of lobes and pinnacles to beige and white color in deeper water or on reef slopes exposed to open ocean conditions. Recent studies (Maragos, 1972) indicate that *M. verrucosa* is

Figure 46.—SCLERACTINIA. *Montipora verrucosa* (Lamarck). A ramose colony showing pinnacles and plates.

resistant to pollution in Kaneohe Bay. In protected environments the coral is able to compete successfully for space against the dominant finger coral *Porites compressa* (Branham and others, 1971; Maragos, 1972). Recently, a parasitic polyclad flatworm, *Prosthiostomum montiporae* Poulter, has been found that feeds on the tissues of *M. verrucosa* (Jokiel and Townsley, 1974). The coral appears to grow rapidly (Maragos, 1972) and may feed on populations of bacteria (Sorokin, 1973).

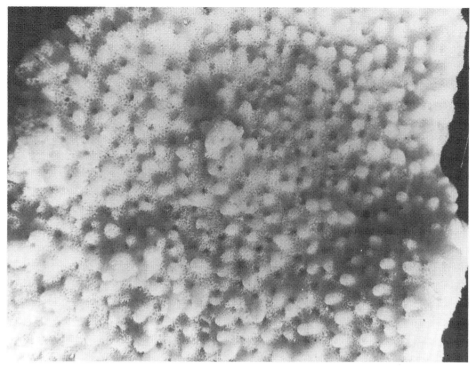

Figure 47.—SCLERACTINIA. *Montipora verrucosa* (Lamarck). Close-up of calices and papillae.

M. verrucosa has been widely reported outside of Hawaii, but the non-Hawaiian forms show subtle but significant differences from the Hawaiian forms. Since the type specimens were originally described from areas outside Hawaii, the Hawaiian forms called *M. verrucosa* may belong to a different variety or species, perhaps the species *Montipora capitata* (Dana 1846). However, resolution of this problem will be difficult since Lamarck's type specimens of *M. verrucosa* have been lost.

Montipora dilatata Studer. This coral is rare and has only been reported on the leeward islands of the Hawaiian archipelago and in Kaneohe Bay, Oahu. The chocolate brown color is similar to that of *M. verrucosa,* which lives in similar environments. As a consequence, *M. dilatata* is not easily recognized. The coral builds horizontally expanding thin sheets, often leaflike, 1.5 cm in thickness (Fig. 48). Sometimes colonies are found to be massive or glomerate. The coenosteum lacks prominent projections. One definitive character of the species is the presence of thin laminar rings or collars surrounding each calyx. Calices are small (0.5 mm in diameter).

Montipora flabellata Studer. This species forms small encrustations that usually assume the shape of the substrate over which it is growing. Most living colonies show a bright turquoise blue color. The surface of the coral is covered by irregular lobes. The calices are small (0.5 mm), evenly spaced, and show only slight development of the septa. Small nipplelike papillae cover the surface and are often fused together into wavy ridges between calices (Fig. 49). The coral is most commonly observed at shallow depths where wave action is heavy. Specimens of a very similar species, *M. venosa,* were reported recently from Kure Atoll (T. F. Dana,

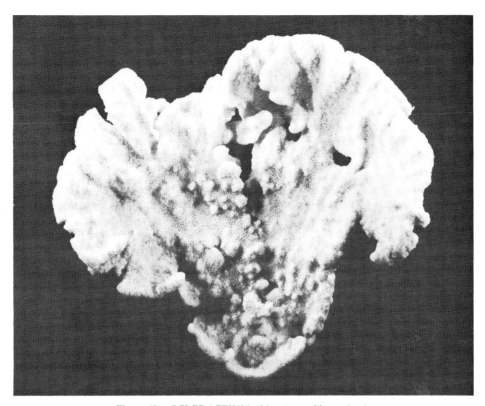

Figure 48.—SCLERACTINIA. *Montipora dilatata* Studer.

1971). These Hawaiian specimens are probably identical to *M. flabellata* (Tom Dana, pers. comm.).

Montipora patula Verrill. This species and another, *M. verrilli,* are extremely difficult to distinguish. Both species show tubercles (rodlike projections) of similar shape and dimension, and both show the characteristic tan color when living. *M. patula* is much more common in Hawaii and can be distinguished by having a ring of tubercles surrounding every calyx (Fig. 50). Calices are small (0.3 mm), and some are elevated while others are depressed. Also *M. patula* shows a foliaceous form with free edges which is lacking for *M. verrilli* (Fig. 51). These differences seem subtle, and in some cases, there are intergrades between the two species. Nevertheless, outside Hawaii, the *M. verrilli* form is common while the *M. patula* form is nearly absent; the opposite is true in Hawaii. Thus, it seems justified that the two forms should be distinguished, at least as separate varieties. It may be necessary to redescribe both species later, since the original descriptions were based on only a few specimens and the definitive characters used for differentiation may not now be considered reliable. *M. patula* is common in Hawaii below depths of 3 m. Some encrusting colonies may measure 2 m or more across, and some foliaceous brackets were reported to extend 50 cm or more in confined environments in Kaneohe Bay. Often encrusting patches of *M. patula* and *M. verrucosa* are seen growing side by side on the reef.

Montipora verrilli Vaughan. As mentioned earlier, this species groups closely with *M. patula,* a fact also confirmed by the numerical taxonomic studies of Powers

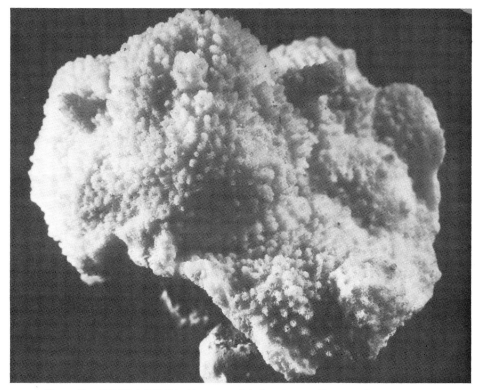

Figure 49.—SCLERACTINIA. *Montipora flabellata* Studer. Close-up of a small colony.

(1970). The calices are larger (0.7 mm to 0.9 mm) on upper surfaces and smaller on under surfaces (0.3 mm to 0.6 mm). A ring of 3 to 6 tubercles often surround the calices (Fig. 52). Contrary to Vaughan's original description, calices seem to be mostly depressed at a uniform level over the surface of the coral. The tubercles all seem to attain the same height above the surface and appear evenly spaced except near calices where they sometimes cluster in rings. Colonies are invariably encrusting in Hawaii and are confined to shallow water, usually on reef flats (Fig. 53).

Family **Agariciidae**

The family Agariciidae in Hawaii is represented by two genera—*Pavona* and *Leptoseris*—and by one subgenus—*P. (Pseudocolumnastraea)*. In general, species of *Pavona* are confined to shallow environments, whereas members of *Leptoseris* are confined to deep water. The five species of *Leptoseris* reported from Hawaiian waters exceed in number those reported elsewhere in the Pacific. This is in part attributed to the considerable amount of deep-water dredging for corals that was conducted near the turn of the century in Hawaii (T. W. Vaughan, 1907). Members of the family Agariciidae in Hawaii are characterized by having well-developed and laminar septo-costae, which are ridgelike extensions of septa outside the boundaries of the calices. In *Leptoseris* the septo-costae are very fine, long, and often curved, whereas those of *Pavona* are thicker, shorter, and straighter. The three species of *Pavona* and the five species of *Leptoseris* are all quite distinct from one another.

Figure 50.—SCLERACTINIA. *Montipora patula* Verrill. Close-up of a portion of colony showing arrangement of tubercles around the calices.

Figure 51.—SCLERACTINIA. *Montipora patula* Verrill.

Pavona varians Verrill. This coral is the most common species of *Pavona* in Hawaii and may occur at depths between 0 m and 25 m or deeper. Most frequently, *P. varians* grows as encrusting lobular masses (Fig. 54) colored tan or bicolored tan and brown when alive. In shady habitats or at depths, colonies may assume a delicate, thin foliaceous form (Fig. 55), brown in color when alive. This latter variety is seen frequently in association with the finger coral *Porites compressa,* where it grows among the bases of the "fingers." Both forms of *Pavona varians* are widely distributed in the islands. The original description of the species by Verrill from Hawaiian specimens is brief and excludes a number of important taxonomic characters. As a result, specimens later described as *P. varians* from reefs outside Hawaii may actually belong to other species. In addition, the skeletal characters of the Hawaiian specimens may merit the placement of this species within the subgenus *Polyastra*. Specimens of *P. varians* are characterized by the presence of elongate angular steep-sided ridges which separate groups of calices located in the valleys (Fig. 56). The septo-costae extend up the sides of the ridges to the crest where they are discontinuous with the septo-costae on the opposite side.

Pavona duerdeni Vaughan. This species is quite distinct from *Pavona varians* in lacking the ridges and in showing a more regular and symmetrical arrangement of the calices (Fig. 57). The growth form in young immature colonies is encrusting or roundly lobate. In larger massive colonies, which sometimes attain heights of 3 m or more, the coral is composed of upright lobes, flattened on the sides, which project outward, resembling disks of 2 cm thickness and 20 cm diameter (Fig. 58). The most common color assumed by living colonies is light gray or pale brownish gray. A small hapalocarcinid crab, *Troglocarcinus crescentus,* lives in crescent-shaped pits in the coral. *P. duerdeni* seems to prefer wave-exposed environments in shallow water far removed from shore. *P. duerdeni* resembles closely *P. clavus,* a species widely distributed in the Pacific. Specimens recently identified as *P. clavus* from Kure Atoll, Hawaiian Islands (T. F. Dana, 1971), probably are synonymous with *P. duerdeni* (John Wells, pers. comm.). *P. explanulata* may also be an earlier name for the Hawaiian specimens referred to as *P. duerdeni.* However, Lamarck's original description of the former is too poor to make a determination, and the latter description of Vaughan's is preferred here.

Pavona (Pseudocolumnastraea) pollicata Wells. This species has been only recently recorded in Hawaii off Kure Atoll (T. F. Dana, 1971), Manele Bay off Lanai, and Kahe and Haleiwa off Oahu. *P. (P.) pollicata* is usually restricted to deeper water below 20 m, but specimens at Kahe were taken from under a ledge at 5 m. Colony form varies from foliaceous to ramose (Figs. 59, 60), with immature specimens assuming an encrusting form. Well-developed colonies recorded elsewhere in the Pacific assume a fingerlike form reminiscent of *Pocillopora* (Fig. 61). However, the branching variety has not yet been reported in Hawaii. Unlike other species of *Pavona,* colonies of *P. (P.) pollicata* show well-differentiated calices raised on pedestals as high as 2 mm. The calices show large centrally located cavities with the septo-costae continuing out and down the calyx walls into the coenosteum. The septa are the typical *Pavona* type. The septo-costae cover the coenosteum, being frequently twisted and bifurcated. The living color is a light brown or gray.

Leptoseris incrustans (Quelch). This species is the most commonly reported *Leptoseris* in Hawaii, probably because of its shallower distribution compared to the remaining species. Colonies of *L. incrustans* in depths of less than 20 m appear

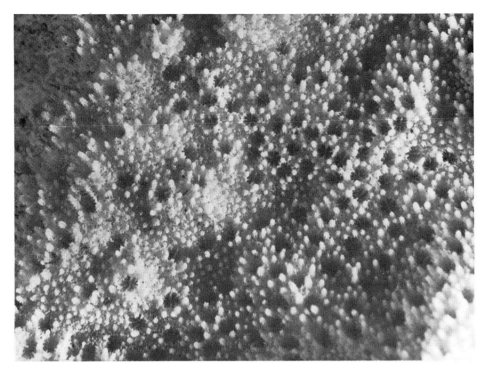

Figure 52.—SCLERACTINIA. *Montipora verrilli* Vaughan. Close-up of a portion of a colony showing calices and tubercles.

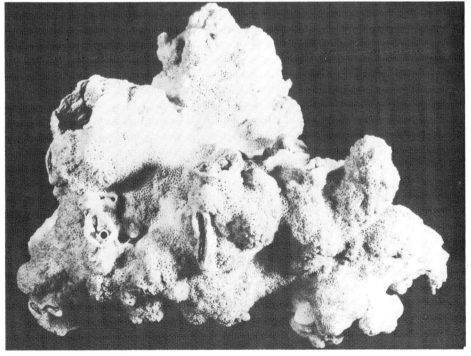

Figure 53.—SCLERACTINIA. *Montipora verrilli* Vaughan.

Figure 54.—SCLERACTINIA. *Pavona varians* Verrill, typical.

Figure 55.—SCLERACTINIA. *Pavona varians* Verrill, foliaceous colony.

Figure 56.—SCLERACTINIA. *Pavona varians* Verrill. Close-up showing calices and septo-costae.

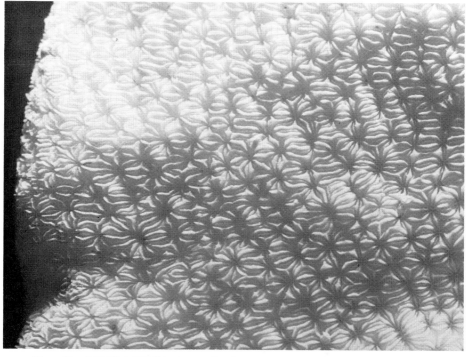

Figure 57.—SCLERACTINIA. *Pavona duerdeni* Vaughan. Close-up of a portion of a colony showing arrangement of calices.

to be restricted to shady crevices or under ledges. One specimen was reported in 1973 by the author at a depth of 4 m on the wall of a boat harbor which was constructed only 3 years previously (Oceanic Foundation, 1975). The colony form is encrusting with irregular swellings (Fig. 62). The characteristic upswelling of the septo-costae between calicular centers is the most definitive character of the species. Colonies rarely exceed 20 cm in diameter. The color is pale brown.

Leptoseris papyracea (Dana) [syn. *L. digitata* Vaughan]. This beautiful flowerlike coral is confined to depths below 25 m and has been reported on several occasions off Oahu and the Kona coast of the island of Hawaii. Young small specimens possess a single calyx. Eventually the free margin of the coral grows and produces several lobes, later developing calices and secondary lobes. In deeper water the lobes appear to elongate into fingers (Fig. 63), but at depths of 30 m or less colonies show only short thickened lobes much like the petals of a flower (Fig. 64). Differences in currents or light conditions may cause these variations in growth form. T.W. Vaughan (1907) originally described the coral as *L. digitata*, but later Wells (1954) synonymized it with earlier described *L. papyracea* (Dana), a widely distributed Indo-Pacific coral.

Leptoseris hawaiiensis Vaughan. The corallum of this species is very thin and the upper surface is flat with a few gentle swellings. Occasionally the margin of the coral becomes upturned, giving *L. hawaiiensis* a cuplike appearance (Fig. 65). The young coral attaches and grows into a thin, funnel-shaped corallum that later extends on all sides in a horizontal plane. The largest colonies are 15 cm or more in diameter. The margin contains few lobes or indentations, the primary basis on which it may be differentiated from the similar coral, *L. papyracea*. The septo-costae of *L. hawaiiensis* are also more compact and less granular than those of another similar species, *L. scabra*. All information on the coral comes from dredgings conducted at the turn of the century. Although the species appears to have a deep bathymetric distribution in Hawaii, it may eventually be reported at shallower depths by SCUBA divers. Many specimens were collected between depths of 50 m to 500 m (T.W. Vaughan, 1907). Those in deeper water may exhibit a more delicate and larger form. Rick Grigg (pers. comm.) has observed large bracketlike colonies of *L. hawaiiensis* at depths of 100 m and more using a deep-diving submersible.

These colonies were pigmented a deep brown color and may have contained the symbiotic algae characteristic of reef corals. However, colonies located at greater depths probably do not contain the algae, if they are present at all.

Leptoseris scabra Vaughan. This species can be easily distinguished from *L. incrustans* by having a free undulating margin and calices centered on the crests of numerous surface swellings (Fig. 66). In contrast, *L. incrustans* lacks free margins and the calices are located in depressions. Unlike other species of *Leptoseris*, *L. scabra* may show enlarged granulated septo-costae on the sides of the swellings or hills. The only specimens from Hawaii have come from deep dredge hauls below 70 m. The species, however, has been reported in shallow water at depths of 25 m at Canton Atoll, Phoenix Islands, located some 2,000 km southwest of Hawaii (Maragos and Jokiel, In press). It seems likely that with increased diving activity *L. scabra* may eventually be recorded in shallower water in Hawaii.

Leptoseris tubulifera Vaughan. This coral is easily distinguished by the cylindrical upright tubes that cover the upper surface. The tubes develop on the edge of the corallum when the growing margin bends around, meets, and fuses. The tubes

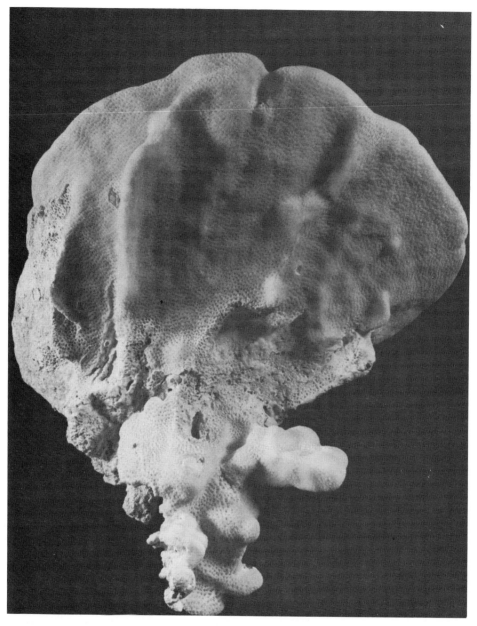

Figure 58.—SCLERACTINIA. *Pavona duerdeni* Vaughan.

then continue to grow upward. Eventually the tubes may bifurcate. The calices appear to be arranged in roughly concentric patterns over the surface. All but one specimen of *L. tubulifera* from Hawaii have been obtained from dredge hauls in deep water (Fig. 67, T.W. Vaughan, 1907). A single small colony (Fig. 68) was recently collected during a SCUBA dive in 30 m of water in the Haleiwa trench off the north shore of Oahu.

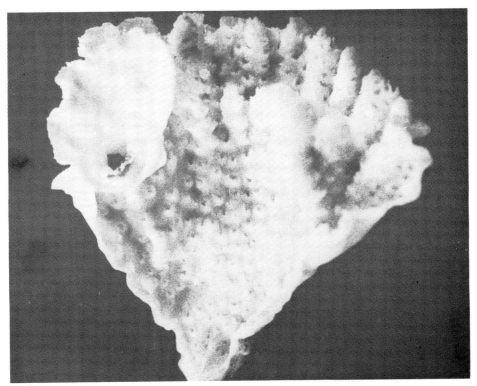

Figure 59.—SCLERACTINIA. *Pavona (Pseudocolumnastraea) pollicata* Wells.

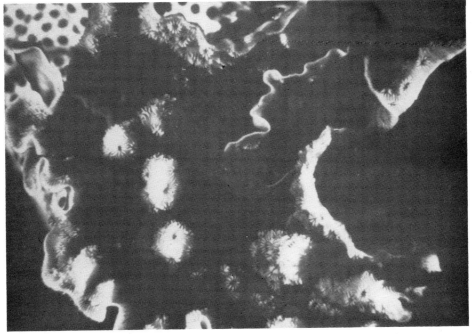

Figure 60.—SCLERACTINIA. *Pavona (Pseudocolumnastraea) pollicata* Wells. Photograph of a living colony.

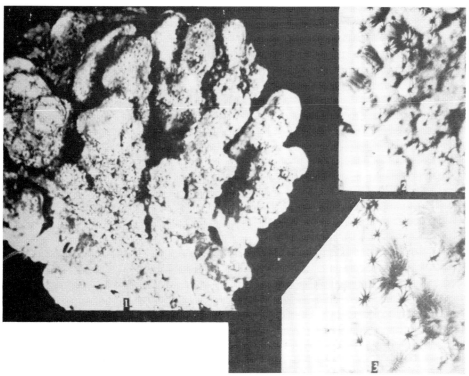

Figure 61.—SCLERACTINIA. *Pavona (Pseudocolumnastraea) pollicata* Wells. A columniform variety of a colony collected from the Marshall Islands. Photograph from Wells (1954).

Figure 62.—SCLERACTINIA. *Leptoseris incrustans* (Quelch).

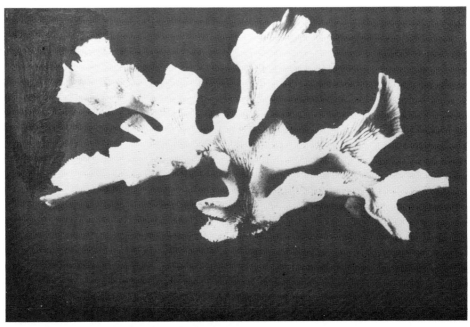

Figure 63.—SCLERACTINIA. *Leptoseris papyracea* (Dana). Deep-water growth form. Photograph from T. W. Vaughan (1907) who described the coral as *L. digitata* Vaughan.

Figure 64.—SCLERACTINIA. *Leptoseris papyracea* (Dana). The more compact form of the species typical of shallower water.

Figure 65.—SCLERACTINIA. *Leptoseris hawaiiensis* Vaughan. Photograph from T.W. Vaughan (1907).

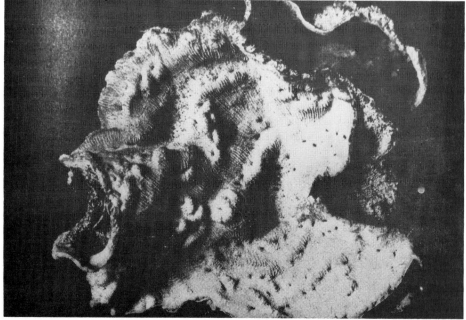

Figure 66.—SCLERACTINIA. *Leptoseris scabra* Vaughan. Photograph from T.W. Vaughan (1907).

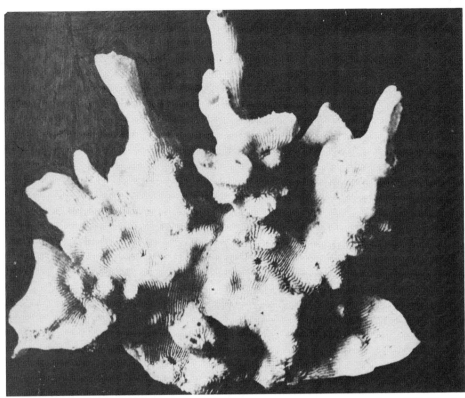

Figure 67.—SCLERACTINIA. *Leptoseris tubulifera* Vaughan. Photograph from T.W. Vaughan (1907). Deep-water specimen.

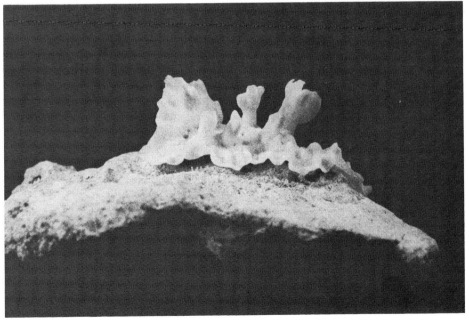

Figure 68.—SCLERACTINIA. *Leptoseris tubulifera* Vaughan. Colony taken from shallow water (30 m).

Family **Balanophyllidae**

Balanophyllia sp. One or more unidentified species of *Balanophyllia*, probably varieties of *B. affinis* (Semper) or *B. hawaiiensis* Vaughan, are known from shallow Hawaiian waters (John Wells, pers. comm.). *Balanophyllia*, like *Tubastraea* (family Dendrophyllidae), is ahermatypic and has tubelike corallites, but it is distinguished from *Tubastraea* by being solitary rather than colonial. However, young individuals frequently attach to dead specimens or to the lower portions of living ones, giving them a colonial appearance (Fig. 69). Most members of the family Balanophyllidae in Hawaii are confined to great depths (Table 3), except the genus *Balanophyllia*, which is also reported on shallow reefs at depths between 10 m and 30 m.

Family **Dendrophyllidae**

Most members of this family are ahermatypic and are generally restricted to the deep, poorly lighted zones of the ocean. The reef-building genus of the family is absent from Hawaii.

Tubastraea coccinea Lesson [syns. *Tubastrea aurea* Quoy and Gaimard and *Dendrophyllia manni* (Verrill)]. This species is the only common Hawaiian ahermatypic coral in shallow water. The coral is composed of small fleshy colonies with the individual calices varying in size, age, and length (Figs. 70a, 70b). The typical clump of 10 to 20 large cylindrical calices may be 5 cm to 10 cm across (Fig. 71). There is considerable variation in the form and structure of the calices. The skeleton is very perforate, the walls of the calices are very thin, porous, and composed of granular ridges. The primary and secondary septa are easily seen, and the upper edge of the septa are well depressed below the surface of the calyx.

The color of living colonies is usually bright orange, but some individuals may appear black. The pigment is not produced by symbiotic algae (zooxanthellae). Because of the preferential absorption of red light near the ocean surface, the "red" colonies at increased depth will appear gray or green under natural light conditions.

Yonge (1940) believes that the species migrated up to the more lighted zones of the ocean during its recent evolution. The coral may occur at a variety of depths and habitats including shady tide pools, the steep walls of ledges, or ceilings of caves. Colonies prefer to grow on elevated surfaces where sediments cannot accumulate and in reduced light where reef corals cannot compete against it for substratum space. Planula larvae of the coral have been reared in the laboratory (Edmondson, 1946a; Harrigan, 1972) and show the same orange coloring of the adults. A predatory nudibranch *Phestella melanobranchia* is specific on the coral and assumes the same orange color and the tentacles of its prey. The cushion star *Culcita novaegineae* has also been observed feeding on *Tubastraea*.

Family **Faviidae**

The family Faviidae contains many genera of reef-building corals, including the brain corals, but in Hawaii only two genera and three species are present. The Hawaiian species rarely ever form large heads; most are small nodular encrustations which grow very slowly. Calices of the faviid corals are usually large and frequently elevated above the coenosteum. All three Hawaiian species grow on the reef platform in shallow water and are able to withstand wide variations in temperature, salinity, wave action, and other factors (Edmondson, 1928).

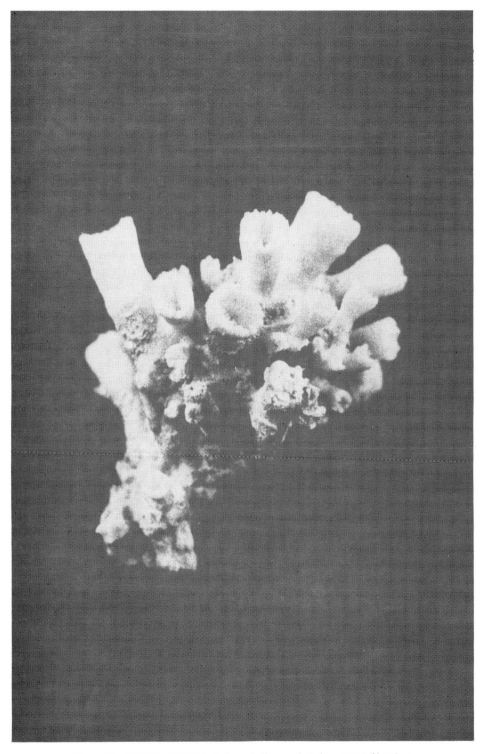

Figure 69.—SCLERACTINIA. *Balanophyllia* sp. cf. *B. hawaiiensis* Vaughan.

Figure 70.—SCLERACTINIA. *Tubastraea coccinea* Lesson: *a*, close-up of extended living polyp; *b*, living colony.

Figure 71.—SCLERACTINIA. *Tubastraea coccinea* Lesson.

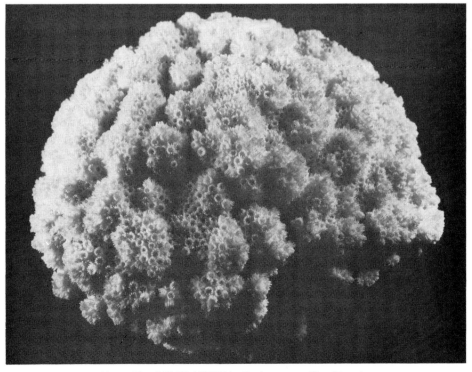

Figure 72.—SCLERACTINIA. *Cyphastrea ocellina* (Dana).

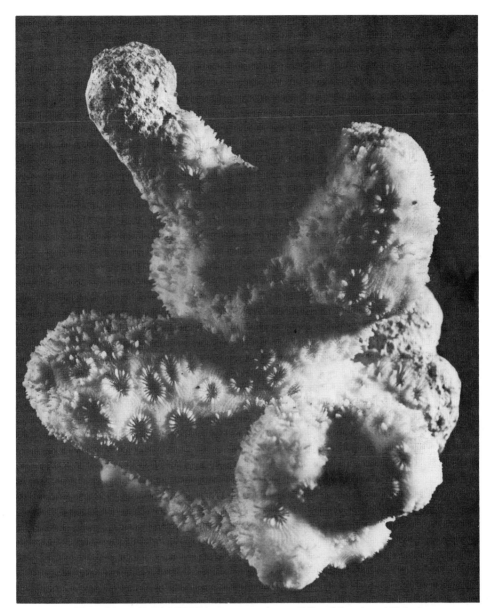

Figure 73.—SCLERACTINIA. *Leptastrea bottae* (Milne-Edwards and Haime).

Cyphastrea ocellina Dana. Colonies of this species form small reddish brown encrustations or clumps 5 cm to 15 cm in diameter. Large, well-developed colonies form mushroom-shaped heads (Fig. 72). The calices average about 1 mm to 5 mm in diameter and are deep. The calyx walls are thick, jagged, and rise vertically above the general colony surface. The septa number 24, and some are sharp and long, frequently projecting above the calyx walls. Both the primary and secondary septa are large, and the tertiaries are small. The calices, orienting randomly in many directions, are usually crowded together. Open areas between the cups are covered with points and plates. This feature and the smaller cups differentiate *Cyphastrea*

Figure 74.—SCLERACTINIA. *Leptastrea bottae* (Milne-Edwards and Haime). Close-up photograph of calices and coenosteum.

from *Leptastrea* in Hawaii. The species is usually confined to shallow water above depths of 8 m and is frequently seen near shore. The coral planulates readily in the laboratory and is ideal for larval study (Edmondson, 1946a). A tiny crab, *Troglocarcinus minutus,* inhabits round pits in the coral.

Leptastrea bottae (Milne-Edwards and Haime) [syns. *L. agassizi* Vaughan and *L. hawaiiensis* Vaughan]. This species forms small nodular encrustations in shallow environments in Hawaii. Calices are nearly round and larger than those of *Cyphastrea* (average 3.5 mm in diameter). A small thin wall surrounds each calyx. Usually the calices are well separated from one another (Fig. 73), but in some colonies the calices are crowded and walls larger (Fig. 74). Calices of *L. bottae* never share common walls as are found in the other Hawaiian species of *Leptastrea.* Characteristic of the species are three complete cycles of septa with the primaries thickest and elevated above the level of the calyx walls. The species is less common than the other Hawaiian faviids. Encrusting patches rarely exceed 15 cm to 20 cm in diameter. The most common habitat of the species is shallow reef flats covered by rubble or coarse sand. The coenosarc tissue (between polyps) is colored olive green to yellow brown when alive, whereas the polyps are usually pigmented a darker brown. In contrast the exsert septal projections look pale or white.

Specimens from outside of Hawaii described as *L. bottae* appear to differ slightly from the Hawaiian forms (John Veron, pers. comm.).

Figure 75.—SCLERACTINIA. *Leptastrea purpurea* Dana.

Leptastrea purpurea (Dana) [syns. *Coelastrea tenuis* Verrill and *Favia hawaiiensis* Vaughan]. Colonies of this species usually occur as thin encrusting sheets over solid surfaces. Some large, well-developed colonies will produce rounded lobes (Fig. 75). The normally large calices of the coral are angular, crowded, irregular, and depressed. Calices average 2 mm to 6 mm in diameter and share common walls. In most forms the funnel-like calices are separated by a narrow groove (Fig. 76), but in some abnormal colonies the groove is nearly absent and the number of septa fewer (Fig. 77). The color of living colonies varies, being pale brown, green, or purple. The central regions of individual polyps are often more strongly pigmented. This hardy coral survives in a variety of habitats from shallow wave-pounded reef flats to deep

Figure 76.—SCLERACTINIA. *Leptastrea purpurea* Dana. Close-up of a typical colony.

reef slopes at depths of more than 50 m. A species of the crab *Troglocarcinus* inhabits deep pits in the surface of *Leptastrea purpurea*.

Family **Fungiidae**

The Hawaiian members of the family Fungiidae (mushroom corals) are all solitary, free-living, and disklike as adults. The immature stage of many fungiids are attached to the substratem, frequently at the end of a stalk, and resemble mushrooms. When the disk portion attains sufficient size, it is broken free and then assumes the adult stage, usually within a year. Breakage from the stalk may be facilitated by the activities of boring algae or by mechanical stress. The flattened disk of the Hawaiian fungiid corals usually contain many elongate septa which radiate from a centrally located mouth depression. Four species belonging to two genera have been reported from Hawaii.

Fungia (Pleuractis) scutaria Lamarck. The solitary coral *F. (P.) scutaria* is the largest and most common of the fungiid corals in Hawaii. The single calyx is elliptical, large (4 cm to 18 cm in diameter), and shows many razor-sharp septa of uniform dimension radiating from the large central mouth depression. The upper margins of all septa show serrated edges. In some specimens the septa continue uninterrupted to the outer disk and are very thin (Fig. 78). In other forms, the septa are curved (Fig. 79, 80), knobby or fused (Fig. 81). The rounded knobby projections, if present, mark the location of some of the many numerous short tentacles that cover the

Figure 77.—SCLERACTINIA. *Leptastrea purpurea* Dana. Close-up showing abnormal crowded and fused calices.

upper surface of active living polyps (Fig. 81). The undersurface is not smooth and is covered with many nipplelike projections. Some adults of the species assume a highly arched form (Fig. 80) which led Döderlein (1901) to describe these as a separate species *(Fungia oahuensis).* The coral survives well under laboratory conditions and has been the focus of frequent physiological and ecological studies. The coral grows slowly compared to the common Hawaiian colonial reef corals (Edmondson, 1929; Bosch, 1967; Maragos, 1972). Some workers (Stephens, 1960, 1962) have maintained that the coral can subsist on dissolved organics. Specimens larger

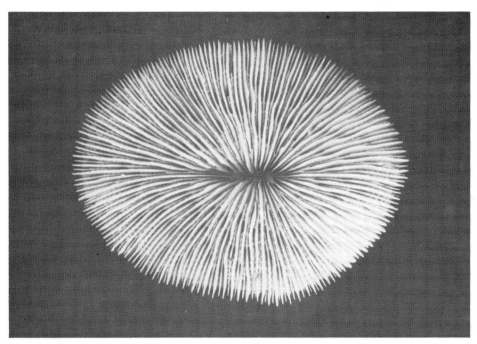

Figure 78.—SCLERACTINIA. *Fungia (Pleuractis) scutaria* Lamarck. Flattened specimen with symmetrically arranged septa.

Figure 79.—SCLERACTINIA. *Fungia (Pleuractis) scutaria* Lamarck, a large robust specimen.

Figure 80.—SCLERACTINIA. *Fungia (Pleuractis) scutaria* Lamarck, an unusual arched specimen originally described as *Fungia oahensis* by Döderlein.

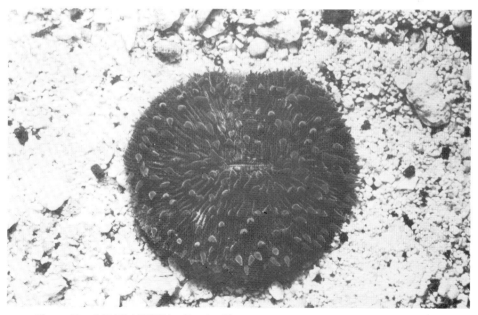

Figure 81.—SCLERACTINIA. *Fungia (Pleuractis) scutaria* Lamarck, live specimen with tentacles extended.

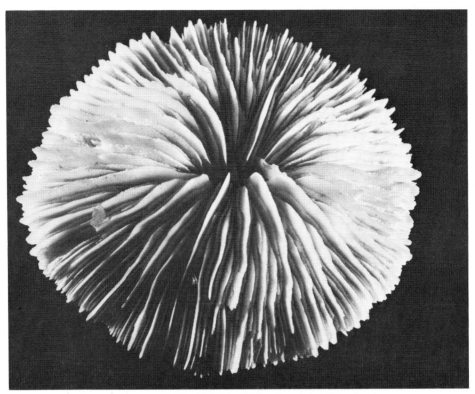

Figure 82.—SCLERACTINIA. *Cycloseris vaughani* (Boschma).

than 18 cm are rarely reported, and Maragos (1972) noted that mortality increases, while both Bosch (1967) and Maragos (1972) noted that growth decelerates with increasing size. Small xanthid crabs take shelter in the concave undersurface of *F. (P.) scutaria,* and a small parasitic gastropod, *Epitonium ulu,* attaches its eggs to the undersurface of the coral (Bosch, 1965). The corals are most common on shallow reef flats, frequently in crevices and depressions, but some specimens have been reported at depths of 25 m or more. Color of live specimens varies from pale brown in the bright sunlight to darker brown in the shade or in deeper water. The huge populations of *F. (P.) scutaria* existing in Kaneohe Bay are beginning to decline, probably because of sewage pollution, collection by divers, or other factors in the bay during recent years. Also the coral appears to grow more slowly where sewage is concentrated (Maragos, 1972). Elsewhere in Hawaii, *F. (P.) scutaria* is commonly associated with mature coral communities protected from wave action. The free-living solitary coral is also hydromechanically adapted to living in environments subject to moderate wave surge (Jokiel and Cowdin, 1976).

Cycloseris vaughani (Boschma) [syn. *Fungia patella* (Ellis and Solander)]. This coral is the most common of the three species of *Cycloseris* present in Hawaii. The coral is small (2 cm to 5 cm), round, and has a convex upper surface, whereas the undersurface is nearly flat and smooth. The septa are thickened at the base and taper to an edge at the upper margin (Fig. 82). The number of septa appear to be proportional to the size of the specimen. *C. vaughani* is capable of movement be-

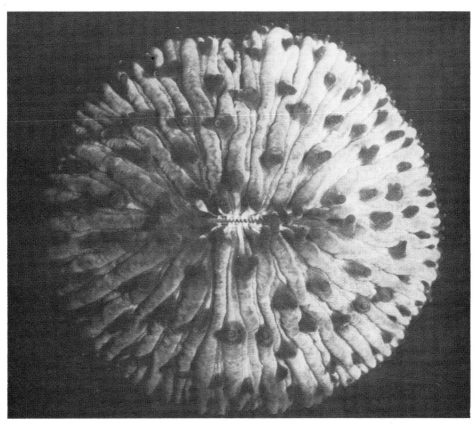

Figure 83.—SCLERACTINIA. *Cycloseris vaughani* (Boschma), live specimen with tentacles extended.

cause of the small size of its skeleton relative to the tentacles and other living tissue (Fig. 83). Usually the coral is confined to pebble flats or hard surfaces at depths greater than 15 m, although some specimens have been seen at depths of 8 m. The coral is rare compared to many other deep-water corals in Hawaii. The species is widely distributed in the Indo-Pacific, and specimens almost identical to those of Hawaii have been collected as far away as the east coast of Africa. Living specimens are usually pale brown or light green.

Cycloseris fragilis (Alcock) [syn. *Fungia fragilis* (Alcock)]. Circular specimens of this species have indented margins or are deeply incised to one side. The circular specimens are composed of semidiscontinuous radial segments, like pieces of pic, which break free under slight mechanical stress (Figs. 84, 85). The segments are dispersed by waves and currents to new environments and eventually regenerate the rest of the disk. Dispersal by fragmentation has also been observed in Hawaii for the related species *F. (P.) scutaria*. Specimens of *C. fragilis* are usually less than 3 cm across. The coral is flat with thin, rounded margins. The undersurface is nearly flat or slightly concave. The septa are very numerous and alternate in thickness and heights. The first and second cycles are more prominent and the margin of the septa are finely serrated. *C. fragilis* is confined to deep reefs in Hawaii below depths of 15 m. The species is widely distributed in the Pacific and Indian Oceans. In Hawaii, specimens are highly gregarious and are often found next to other species of

Figure 84.—SCLERACTINIA. *Cycloseris fragilis* (Alcock).

Figure 85.—SCLERACTINIA. *Cycloseris fragilis* (Alcock), live specimens.

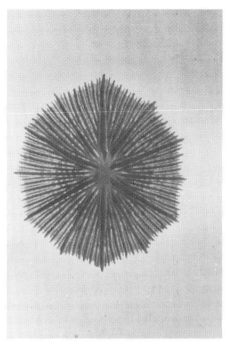

Figure 86.—SCLERACTINIA. *Cycloseris hexagonalis* (Milne-Edwards and Haime), the skeletal specimen is backlighted to show septal arrangements.

Figure 87.—SCLERACTINIA. *Cycloseris hexagonalis* (Milne-Edwards and Haime), live specimen.

Cycloseris. Further study is required on the variation and ecology of *Cycloseris* before the validity of many separately described species in this genus can be documented.

Cycloseris hexagonalis Milne-Edwards and Haime. This species is the smallest and rarest of the fungiid corals found in Hawaiian waters. The corallum is angular with sharp margins and rarely exceeds 2 cm in diameter. The first cycle of septa are larger and more prominent than the others (Fig. 86). The coral has been reported off Kaneohe Bay, off Waikiki in 15 m, and once off Makaha at 11 m depth. The generally small specimens of *C. hexagonalis* are probably confined to deeper water where individuals will not be tossed about by wave surge and strong tidal currents. Specimens of *C. hexagonalis* are usually found in groups of several individuals or more. Living specimens (Fig. 87) are usually brown, but occasionally phosphoresce a yellow orange color in deeper water.

Large numbers of the skeletons and live specimens of *C. fragilis* and *C. hexagonalis* have recently been reported at several locations in Hawaii at depths of 400 m or more (Neighbor Island Consultants, Inc., 1977).

Family **Pocilloporidae**

The family Pocilloporidae contains the important Pacific reef coral genera *Seriatopora, Stylophora,* and *Pocillopora,* but only the last is represented in the Hawaiian Islands. At least nine species of *Pocillopora* have been reported or described from Hawaii, but only five are recognized herein as good species. The systematic study of *Pocillopora* is made difficult because of the lack of septal development and other structures of the calices, thus placing more emphasis on growth-form characters which are extremely variable and unreliable within individual "species" of *Pocillopora*. Even among the five species recognized here there are intermediates which are difficult to place within single species. All Hawaiian species of *Pocillopora* are branching (ramose or cespitose) to some extent, and all seem to grow to definite size and shape on the reefs. Characteristic of most species are wartlike projections, termed verrucae (Fig. 88) upon which are crowded aggregations of calices. The calices are deep, tiny, and jagged polygons which show only rudimentary septa. Most species lack a columella in the calices. One of several species of the crab *Trapezia* lives commensally near the base of the branches of each species of *Pocillopora* (Barry, 1965). Several of the corals, most notably *P. damicornis,* also develop galls which enclose the gall crab *Hapalocarcinus marsupialis*. Recently an undescribed species of *Pseudocryptochirus* has been found in chambers on the branches of *Pocillopora eydouxi* and *P. meandrina* (John C. McCain and Steve L. Coles, pers. comm.).

Pocillopora damicornis (Linnaeus) [syn. *P. cespitosa* Dana]. This species is found widely distributed in the Indo-Pacific region from Panama to the Red Sea. There are a number of synonyms for the species and the Hawaiian specimens described as *P. brevicornis* probably also belong to *P. damicornis* (Table 2). Typical colonies are finely branched and are found in small bushy clumps up to 15 cm in diameter. The variation was termed "bewildering" by T.W. Vaughan (1907), who described three separate varieties. The branchlets are long and slender for specimens (Fig. 89) in quiet or deeper water, but may be more robust (Fig. 90) where wave action is common in shallow water. The branches themselves (Fig. 91) could be considered as elongated extensions of the verrucae which typify other species of

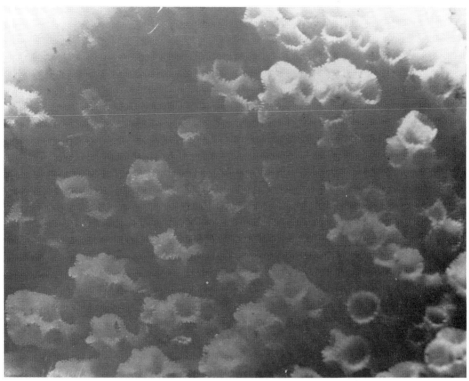

Figure 88.—SCLERACTINIA. Close-up of *Pocillopora* showing arrangement of calices and verrucae.

Figure 89.—SCLERACTINIA. *Pocillopora damicornis* (Linnaeus), typical.

Figure 90.—SCLERACTINIA. *Pocillopora damicornis* (Linnaeus), robust form characteristic of shallow wave-exposed environments.

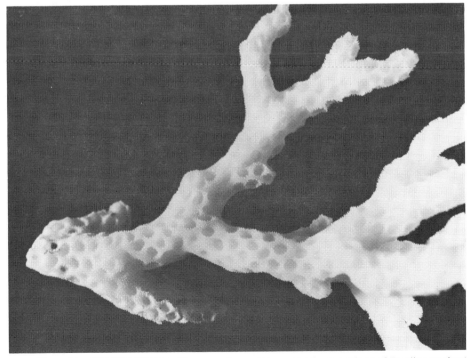

Figure 91.—SCLERACTINIA. Close-up of a small branchlet from a colony of *Pocillopora damicornis* (Linnaeus).

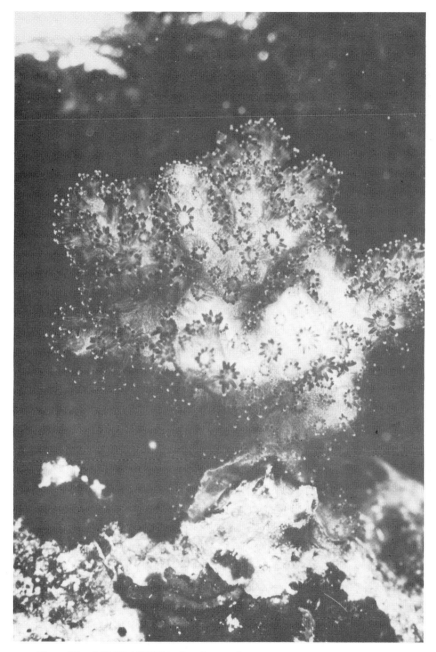

Figure 92.—SCLERACTINIA. *Pocillopora damicornis* (Linnaeus), live specimen.

Pocillopora. The branching begins at the encrusting base, giving rise to many branchlets. In most specimens the septa are poorly developed and the columella nearly absent. At the termini of the branches, the cups are crowded together with little distance separating the polyps (Fig. 92). The calices range from 0.4 mm to 1.0 mm in diameter and are variable in shape. The walls of the terminal calices sometimes flare outward. The coral is most commonly reported within protected bays or

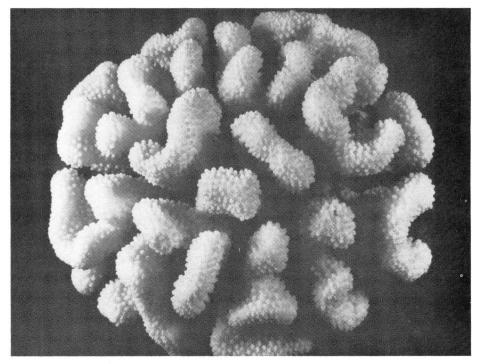

Figure 93.—SCLERACTINIA. *Pocillopora meandrina* Dana var *nobilis* Verrill, typical.

Figure 94.—SCLERACTINIA. *Pocillopora meandrina* Dana var *nobilis* Verrill, specimen showing greater branching tendency and longer verrucae.

upon the inner portions of large reef flats away from breaking waves. The coral will also occur where there is wave action as long as loose substratum material that could damage the fragile colonies is absent. Specimens of *P. damicornis* are rarely reported below depths of 10 m and seem to be strongly light-dependent (Maragos, 1972). Other studies indicate that the species is one of the earliest coral colonizers on reef substrates. Edmondson (1929) hypothesizes that colonies live for only 6 or 7 years. Edmondson, as well as Harrigan (1972), studied planulation in the coral and noted the process to be synchronized to phases of the moon and other factors. Colonies in shallow water are colored light brown, and living specimens are colored darker brown in deeper water.

Pocillopora meandrina Dana var. *nobilis* Verrill. This species is the most common *Pocillopora* in Hawaii and occurs at depths ranging from 0 m to 30 m or more. It is the dominant coral on hard substrates in shallow water (less than 3 m depth). Studies by Grigg and Maragos (1974) indicate this species to be the earliest coral colonizer on submerged lava flows that have entered the ocean off the island of Hawaii during recent years. Commonly called "rose coral," colonies of *P. meandrina* are characterized by heavy leaflike branches often forked near the ends and showing nearly equal length. In some forms (Fig. 93) the verrucae are regular and the spaces between the branches are of uniform width and meander. In other varieties the verrucae are more irregular (Fig. 94) while in other forms the verrucae are so elongated that the specimens resemble *P. damicornis*. The calices are irregular in shape and size and lack development of the septa, columella, and other structures. The average size of calices is 0.7 mm in diameter. The species prefers wave-agitated environments and may be absent from protected regions such as inner Kaneohe Bay. *P. meandrina* appears to be strongly light dependent and is sensitive to sediment pollution (Maragos, 1972). Recently the coral was reported in tide pools located above low tide level along the rocky coast near Makena, Maui. The coral reaches maximum size at 30 cm to 50 cm in diameter, after which colonies stop growing or eventually die. The species prefers sediment-free environments in shallow water which are not dominated by *Porites,* a coral that seems to outcompete *Pocillopora* for substratum cover in deeper protected environments. The color of living colonies varies from brown to chartreuse to pink.

Pocillopora ligulata Dana. Colonies of *P. ligulata* superficially resemble those of *P. meandrina,* but there are a number of conspicuous differences. The more delicate appearance of *P. ligulata* is a reflection of the thinner and more elongate branches which join together at the bases to form wavy folds (Fig. 95). Verrucae are not as prominent in this species and may be nearly lacking at the basal portions or branches. Also many calices of individual colonies of *P. ligulata* show a distinct columella. The pigmentation of living specimens of the coral is different, being a light golden brown. Colonies commonly range between 15 cm and 30 cm in diameter, with some specimens as large as 45 cm. The branches are not of equal lengths, giving colonies an irregular outline, and the ends of branches do not twist and fork as they do in *P. meandrina*. The calices have thin walls, and the interior of the cups are cone-shaped. This irregular branching has led to a number of synonyms of *P. ligulata,* listed in Table 2. Colonies are found on the reef in shallow water, sometimes near heavy wave action, usually far from shore. The coral is widely distributed in the tropical Pacific Ocean.

Pocillopora molokensis Vaughan. Colonies of *P. molokensis* are not common and do not form the branching hemispherical clumps typical of the other Hawaiian

Figure 95.—SCLERACTINIA. *Pocillopora ligulata* Dana.

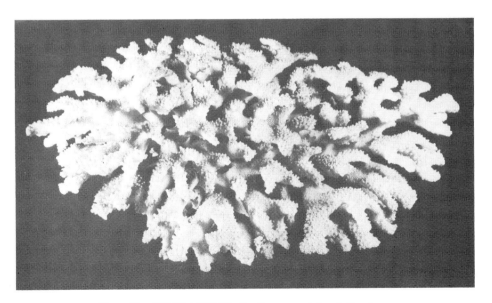

Figure 96.—SCLERACTINIA. *Pocillopora molokensis* Vaughan.

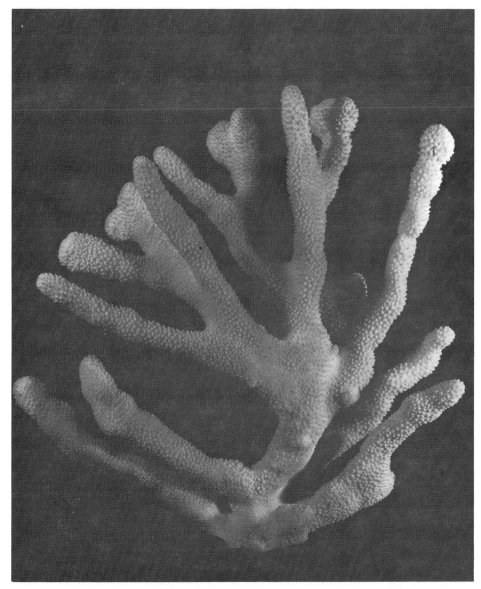

Figure 97.—SCLERACTINIA. *Pocillopora eydouxi* Milne-Edwards and Haime.

members of the genus. Instead, colony shape appears more flattened because of the predominantly horizontal direction of growth by the branches (Fig. 96). Branches are 1 cm to 5 cm thick, have many divisions, and are not flattened in a vertical dimension as with *P. meandrina*. Wartlike verrucae are absent from the underside of the basal branches. The branches decrease in diameter from base to tip, and the tips are usually pointed. The calices have diameters that vary between 0.7 mm and 1.3 mm, are shallow, and lack development of the septa. Colonies range from 15 cm to 50 cm in diameter, with the larger colonies being found in deeper water. *P. molokensis* is among the deepest-living members of *Pocillopora* in Hawaii and has been found at depths of 40 m and 50 m off Molokai and Lahaina, Maui, respectively. The

Figure 98.—SCLERACTINIA. *Pocillopora eydouxi* Milne-Edwards and Haime; living specimen surrounded by platforms of the finger coral *Porites compressa* Dana.

coral has also been reported in shallow water (5 m) off Kaneohe Bay, Oahu.

Pocillopora eydouxi Milne-Edwards and Haime [syn. *P. modumanensis* Vaughan]. This species is the largest and most striking *Pocillopora* in Hawaii. On some reefs colonies may reach heights of 80 cm or more. The thick pipelike branches of *P. eydouxi* are cylindrical and lack many divisions, growth being predominantly vertical (Fig. 97). From a distance large colonies resemble the blunt antlers of a moose. In deeper water the branches are widely spaced and elongated, but in shallow water the branches are stunted and are closer together. The tips of some of the branches may be flattened (Fig. 98). The terminal flattening of the branches of the smaller shallow-water colonies may result in confusing this species with *P. meandrina*. However, calices of *P. eydouxi* possess distinctly developed septa and a styliform columella, which are absent from *P. meandrina*. The verrucae are taller and the calices more crowded in *P. eydouxi;* the calices average 0.7 mm to 1.0 mm in diameter. Schools of the damselfish *Dascyllus* and other species are often seen hovering near large colonies of *P. eydouxi*. Observations have also shown that the coral can withstand heavy wave action. Frequently reported are colonies that have been "uprooted" and carried shoreward by waves, but which still survive. In Kaneohe Bay a detached colony was found to have regenerated its damaged branches and was living 2 km shoreward from its common habitat (Maragos, 1972). *P. eydouxi* is a deep-water coral and colonies have been reported at depths between 12 m and 40 m in Hawaii. The color of living colonies is brown and usually darker than the other species of the genus. The species is widely distributed in the Indo-Pacific.

Family **Poritidae**

The family Poritidae is represented by the important genus *Porites* which is the most common and widespread genus of hermatypic corals in Hawaii and in many other tropical regions. In Hawaii the genus *Porites* contributes more varieties and species of corals than any other and more mass to the reefs than all other corals combined. More than 18 species and 40 varieties have been recognized but many of these are rare, obscure, or difficult to differentiate from one another. This report recognizes only 10 common species, most of which can easily be distinguished in the field. The most unifying feature of the genus *Porites* are the polygonal-shaped calices which often resemble frosted snowflakes because of the complex development of the septa and other structures of the calices.

Porites compressa Dana. Commonly called finger coral, *Porites compressa* is one of the two most common species of *Porites* in Hawaii. At least 16 forms and 2 subforms have been described for this extremely variable species, but all show, to some degree, the characteristic fingerlike branching and porous skeletons. In some forms the fingers may be elongate and show divisions (Fig. 99). In other varieties adjacent branches may be fused into long wavy folds. Only immature colonies still too young and small to send up projections will lack the characteristic fingers. The tips of the branches are usually blunt or flattened, and the branches themselves are cylindrical, show little tapering, and are frequently bifurcate or trifurcate. Calices of this species are typical of *Porites*, 1.5 mm to 2.0 mm in diameter, and are nearly flush with the surface (Fig. 100). The upper surfaces of the septa contain elevated rods, giving the calyx a snowflakelike appearance. Adjacent colonies of the species often fuse or crowd together and form continuous platforms, some of which can cover many square meters of the reef. The coral has been reported from low tide level to depths of 50 m. This coral is most common in wave-protected waters, especially in bays or on deeper reef slopes on the leeward sides of the islands. Because of its rapid growth rate and peculiar growth form, *Porites compressa* outcompetes other corals for space. The coral comprises 85 percent of the total coral population in Kaneohe Bay (Maragos, 1972) and dominates deeper reef slopes off the Kona coast of Hawaii Island and in other protected environments (Dollar, 1975). However,

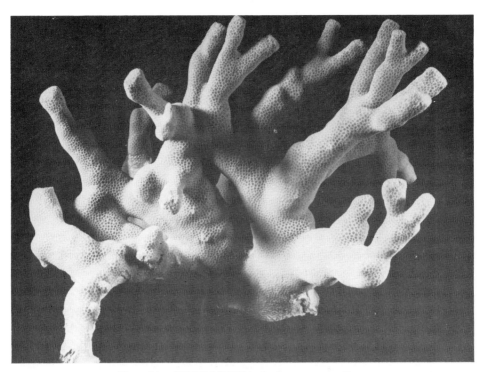

Figure 99.—SCLERACTINIA. *Porites compressa* Dana.

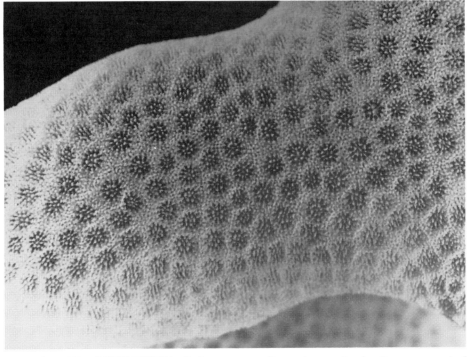

Figure 100.—SCLERACTINIA. *Porites compressa* Dana; close-up showing the arrangement of calices on a fingerlike projection of a colony.

Figure 101.—SCLERACTINIA. *Porites duerdeni* Vaughan, type specimen.

because of its fragile nature and sensitivity to abrasion, *P. compressa* is seldom reported where wave action is heavy. Other studies have shown the coral to be sensitive to sewage and sediment pollution. *P. compressa* also contributes considerable bulk to fossil reefs. The city of Honolulu, including Waikiki, rests upon an uplifted fossil coral reef composed largely of the skeletal remains of *P. compressa* (Pollock, 1928). Although finger coral belonging to one or more of several ramose species of *Porites* is extremely common in Hawaii, Samoa, and other tropical areas, it has not been reported from the Line Islands and Phoenix Islands, which are both close to Hawaii.

Porites duerdeni Vaughan. This is another branching species of *Porites* whose growth form is similar to that of *P. compressa*. Although the structure and size of the calices of *P. duerdeni* are readily different from that of *P. compressa* (compare Figs. 99 and 101) the former is hard to distinguish in the field and is hardly ever reported. The calices of *P. duerdeni* are deep, funnel-like, and vary between 1.0 mm and 2.5 mm in diameter, and are separated by a thin continuous elevated wall. The septa are somewhat thick, and the upper edges are broken into a number of starlike points of fairly regular arrangement. Technically the septa of *P. duerdeni* are composed of inwardly inclined trabeculae, showing a large angle of divergence with respect to the wall compared to *P. compressa*, where the angle is much less divergent (T.W. Vaughan, 1907). The columella is large and conspicuous in the cups. The branches of *P. duerdeni* are thick and show numerous constrictions and swellings and are often fused to adjacent branches. Other varieties of *P. duerdeni* show swollen lobes at the branch tips. The branching angle is wide and almost oblique com-

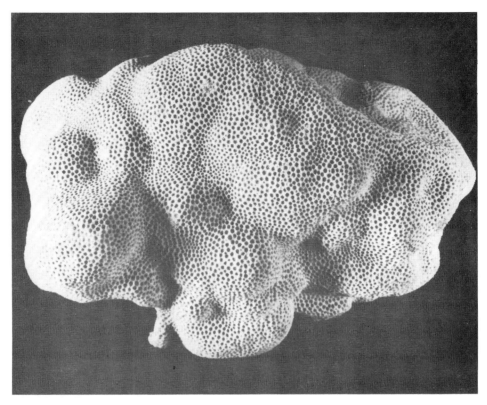

Figure 102.—SCLERACTINIA. *Porites lobata* Dana, typical.

pared to *P. compressa*. Some colonies may be large, as much as 60 cm across. *P. duerdeni*, because of its similarity to common finger coral, has only been reported in Kaneohe Bay, Oahu, and Ahihi Bay, Maui. Colonies are brown when alive.

Porites lobata Dana. *P. lobata* is perhaps the most abundant and widespread of reef corals in the Hawaiian Islands. The coral is quite variable in shape and no less than 5 forms and 5 subforms have been described. This coral most commonly assumes an encrusting or massive form on reefs. Frequently the edges of encrusting colonies are upturned and long narrow cracks are produced by commensal alpheid shrimps (R.A. Vaughan, 1973). Other varieties of *P. lobata* include erect columns, smooth lobes, or hemispherical lobes (Fig. 102). The walls of the calices are membranelike and elevated into sharp edges (Fig. 103). The columella structure is usually compact. Calices on the sides of colonies are very shallow but are more funnel-like on upper surfaces. As with other species of common *Porites*, the calices show an angular outline and a frosted snowflake appearance. The diameter of the calices varies from 1.5 mm to 2.0 mm. Despite the generally dissimilar forms of *P. compressa* and *P. lobata*, the species are so common that a number of "intermediate" forms do exist which can make identification difficult. In *P. compressa* projections are either fingerlike, or at least one dimension of the projections is never thicker than a finger (2 cm). In contrast, colonies of *P. lobata* may show prominent projections on occasion, but these are never thin or fingerlike. Most commonly projections are domes, inverted cones, columns, or irregular pinnacles. Living colonies of *P. lobata*

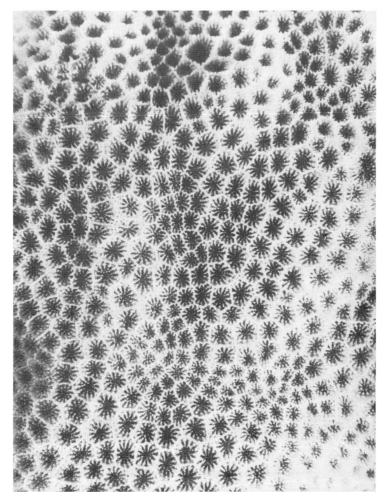

Figure 103.—SCLERACTINIA. *Porites lobata* Dana, close-up showing arrangement of calices.

are olive green, chartreuse green, brown, and sometimes blue. The coral has been reported from the intertidal zone to depths greater than 60 m, but is most common in wave-exposed reef slopes between depths of 3 m and 15 m. Colonies sometimes attain heights and diameters of several meters or more.

Porites evermanni Vaughan. Superficially the massive colonies of *P. evermanni* resemble those of the more common *P. lobata,* but there are gross differences in the structure of the calices. The septa are made up of thin plates instead of the more granular protrusions of other Hawaiian *Porites*. In addition, the septa of *P. evermanni* show an arrangement of vertical points on their upper surfaces that gives the calices a double-walled appearance (Fig. 104). Calices are all shallow and vary from 1.0 mm to 1.5 mm in diameter. Colonies of *P. evermanni* assume flat-topped or cubelike lobes with rounded edges in shallow water (Fig. 105). Living colonies are frequently large and show gray, brown, or purple coloration. The coral is most common on exposed reef flats at depths above 10 m.

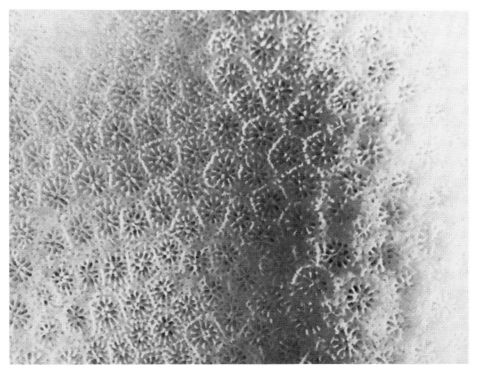

Figure 104.—SCLERACTINIA. *Porites evermanni* Vaughan; close-up showing arrangement of calices.

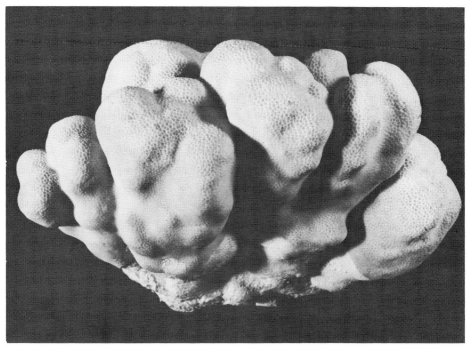

Figure 105.—SCLERACTINIA. *Porites evermanni* Vaughan.

 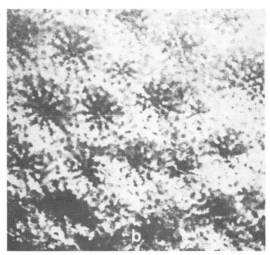

Figure 106.—SCLERACTINIA. *Porites studeri* Vaughan; *a*, small colony; *b*, close-up of calices. Photograph taken from T. W. Vaughan, 1907.

Porites studeri Vaughan. This is the only Hawaiian species of *Porites* apparently restricted to deep water. The structure of the calices and the form of the colonies are distinct from other species of the genus. The most representative form is nearly spherical (Fig. 106*a*), ranging between 20 mm and 30 mm. Calices are polygonal and vary between 1.5 mm and 2.0 mm in diameter. Calices are excavated but shallow (Fig. 106*b*). Colonies of *P. studeri* can be distinguished from those of *P. lobata* by their small size at maturity, subspheroidal form, and shallow calices. In addition, the septa of *P. studeri* are thicker, slope more gradually to the bottom of the calices, and result in narrower interseptal spaces. The columella is not situated in a pronounced depression, as in *P. lobata*. All information on *P. studeri* comes from deep-water dredgings (28 fathoms to 43 fathoms), conducted at the turn of the century (T. W. Vaughan, 1907), and the species has yet to be collected during SCUBA dives. However, the coral has been reported in the Marshall Islands at shallower diving depths.

Porites pukoensis Vaughan. Specimens of *P. pukoensis* are found as massive colonies with conical projections on shallow reefs in Hawaii. It is not a common coral and may be easily confused with specimens of *P. lobata,* since the latter also show conical projections in some forms. *P. pukoensis* is composed of cone-shaped columns with pointed summits (Fig. 107). In contrast, colonies of *P. lobata* have the cones wider and shorter, and frequently elongated into long ridges. The characters of the calices are of greater taxonomic importance in distinguishing *P. pukoensis*. Calices are polygonal, composed of 5 unequal sides, deep, excavated, and they vary in size between 1.25 mm and 1.50 mm in diameter. The upper edges of the calical walls are made up of a number of vertical points, and 3 vertical starlike points are found on the upper edge of each septum. When viewed from the side, the calyx walls appear elevated and the septa recessed. T. W. Vaughan (1907) indicates the species may be an intermediate form between a *Porites lobata* and *P. compressa*. Specimens of *P. pukoensis* in Hawaii are usually not larger than one meter in di-

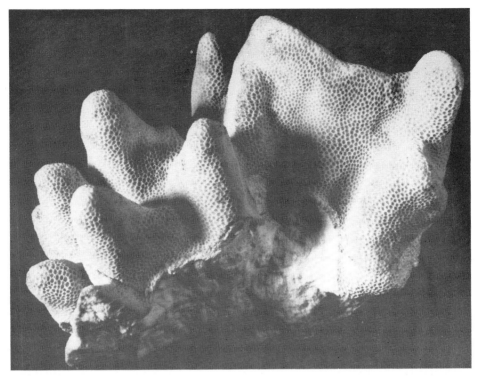

Figure 107.—SCLERACTINIA. *Porites pukoensis* Vaughan, type specimen.

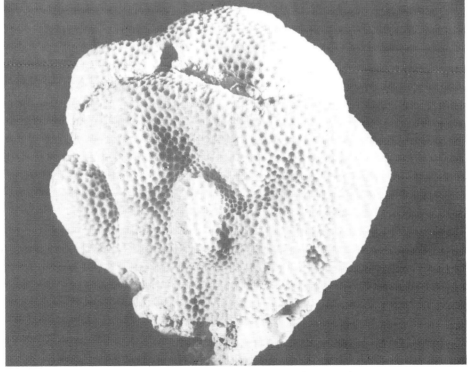

Figure 108.—SCLERACTINIA. *Porites brighami* Vaughan.

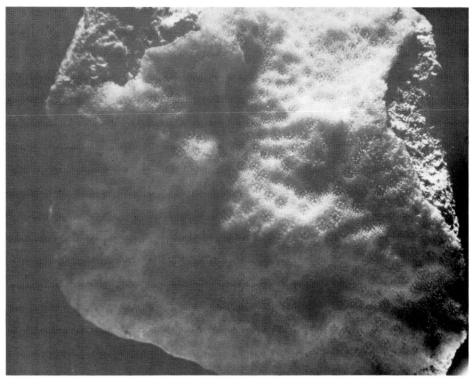

Figure 109.—SCLERACTINIA. The type specimen of *Porites (Synaraea) hawaiiensis* Vaughan, which is considered in this report to be a young colony of *P. (S.) convexa* Verrill.

ameter. Colonies matching Vaughan's description of the species have been reported at Fanning Atoll (Line Islands), 1,600 km south of Hawaii (Maragos, 1974).

Porites lichen Dana. This uncommon species of *Porites* forms small encrustations in shallow water and some rare varieties show free edges for 1 cm, the undersurface appearing very smooth. The most striking feature of the species is the tiny shallow calices which are tightly packed and are often prominent in minute thin ridges that give the surface a reticulate appearance. Color of living colonies is a light yellow to yellow green. The species has been reported from reef flats in Kaneohe Bay and Kailua Bay, Oahu, and elsewhere. This species is widely distributed in the Pacific but has only been recently reported from Hawaii. Quelch (1886) also described some specimens from Hawaii as *P. lichen*, but later T.W. Vaughan (1907) determined these to be young specimens of *P. lobata*.

Porites brighami Vaughan. This species occurs as small lobed or knobby encrustations on shallow hard substrate on wave-washed reefs. The large funnel-like calices are the species' most striking character (Fig. 108). The cups are deep and the septa do not extend more than one third of the way to the center of the calices. The septa also appear to be thickened at the distal ends. In contrast to other species of *Porites*, the columella within *P. brighami* is so deep as to be nearly unrecognizable. The diameters of the cups range from 1 mm to 2 mm. Colonies of *P. brighami* can be easily differentiated from the more common *P. lobata* by the simpler, deeper, and more funnel-like calices. Living colonies are gray, light blue, and sometimes yellow. *P. brighami* has been reported on all major Hawaiian islands at depths less than 10 m.

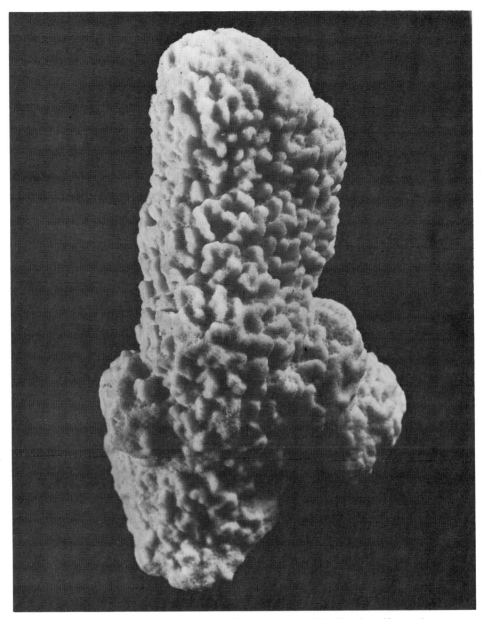

Figure 110.—SCLERACTINIA. *Porites (Synaraea) convexa* Verrill; columniform colony.

Porites (Synaraea). The subgenus *P. (Synaraea)* is well represented in Hawaii and may be differentiated from the other *Porites* by its small, round, and shallow calices which are usually separated from one another by distances as great or greater than the diameters of the calices themselves. Because of the tiny nature of the calices, representatives of *P. (Synaraea)* can be easily misidentified as members of the genus *Montipora,* but close examination will reveal the characteristic "snowflake" calices of *Porites* as opposed to the starlike calices of *Montipora.* T.W. Vaughan (1907) described *P. (S.) hawaiiensis* (Fig. 109), but the present author

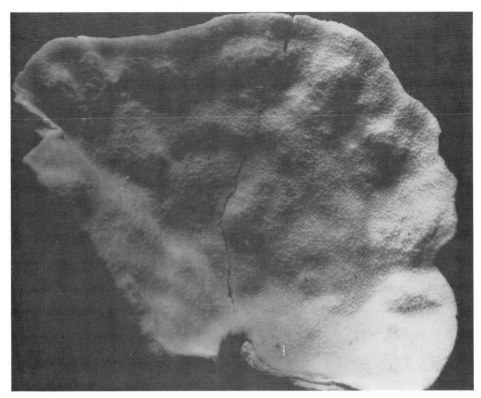

Figure 111.—SCLERACTINIA. *Porites (Synaraea) convexa* Verrill; foliaceous colony.

feels that the type specimen was too small and perhaps immature to merit separate species rank. Furthermore, the original description of the species was based on only a single specimen. I contend that this specimen probably belongs to *P. (S.) convexa*, which had not yet been reported in Hawaii at the time of Vaughan's study.

Porites (Synaraea) convexa Verrill. This species has only recently been reported from Hawaii, but is widespread elsewhere in the tropical Pacific. In Hawaii the coral has been reported only off the west coasts of Maui and Hawaii. The growth form is somewhat variable, but the massive form with cylindrical lobes (Fig. 110) and the foliaceous platelike form (Fig. 111) are the most common. The latter form occurs in deep water or in crevices and other environments where light intensity is reduced. Some of the semicircular plates of the foliaceous variety may be very thin (less than 1 cm thick) and fragile. In the massive form the calices are crowded between irregularly elongated ridges that rise above the general level of the coral surface, giving the colony a rough appearance. These ridges are reduced or absent in the foliaceous form. The color of the living colonies is blackish brown.

Porites (Synaraea) irregularis Verrill. This species is easily distinguished from *P. (S.) convexa* in showing a tendency toward branching. The tips of the branches are nodular, or appear roughly eroded and fused to adjacent branches (Fig. 112). This feature gives the species the appearance of a cauliflower. The calices are larger in this species compared to the other. The branch tips are yellow, white, or gray, whereas the lower portions are colored deep amber when living. *P. (S.) irregularis* is not common in Hawaii, although it has been reported at several reefs including

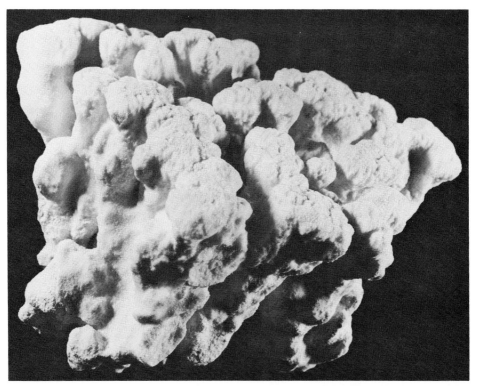

Figure 112.—SCLERACTINIA. *Porites (Synaraea) irregularis* Verrill.

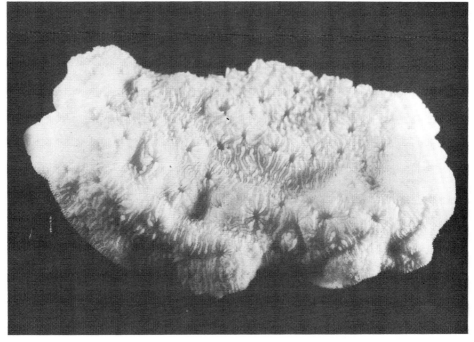

Figure 113.—SCLERACTINIA. *Coscinaraea ostreaeformis* Van der Horst.

Hanauma Bay and Kaneohe Bay, Oahu; Pukoo, Molokai; and on the south side of Kahoolawe Island.

Family **Siderastreidae**

Coscinaraea ostreaeformis Van der Horst. This species was only recently reported in Hawaii during SCUBA dives in deeper water (Maragos, 1972). The colony form is usually encrusting, but some well-developed specimens may send out laterally expanding brackets. Colonies are small and are rarely more than 20 cm across. The calices and septo-costae form beautiful intricate patterns over the surface of the corallum (Fig. 113). Calices show well-defined cavities, and the septa extend outward as well-developed costae, radiating out from the calical swellings and alternating in height. The septa and costae are coarsely granulated and are similar to these characters found in *Psammocora,* except they are on a much larger scale. The septa are numerous, numbering 12 to 14 or more, and reach the columella where they show equal thickness. The columella is a distinct papillalike projection about 0.8 mm in diameter. The calices are normally 4 mm to 5 mm apart, but become more spread out in explanate colonies. Living colonies are pale tan or nearly white in color. In Hawaii the coral is most common on rocky walls or ledges, usually at depths below 20 m. The species has been reported several times on other reef areas in the tropical Pacific, but is not common.

The Hawaiian specimens of *C. ostreaeformis* are identical to those figured in Wells (1954) from the Marshall Islands but appear different from the figured type specimens in Van der Horst (1922). Thus, it is possible that specimens called *C. ostreaeformis* from Hawaii may belong to a separate variety or species of *Coscinaraea.*

Family **Thamastreidae**

Corals of this family have large shallow calices with finely granulated septa. The common members of *Psammocora* grow as many-branched, hemispherical clumps as large as 10 cm in diameter, or as encrustations up to 1 m across. All Hawaiian species are found in shallow water and at least two species have been reported in deeper water. The genus *Psammocora* is represented by two subgenera and three species in Hawaii. The subgenus *(Stephanaria)* is supposedly distinguished from *Psammocora* by the possession of papilliform pali and columella.

Psammocora (Stephanaria) stellata (Verrill) [syns. *Stephanaria stellata* (Verrill) and *P. (S.) brighami* (Vaughan)]. *P. (S.) stellata* is the most common Hawaiian member of the family. On shallow reef flats the species is variable in size (5 cm to 10 cm in diameter). Colonies are frequently unattached to the substrate. The branches are of unequal length and are commonly bifurcate. In some branching forms the branches are close together, contorted, and stubby (Fig. 114), whereas in others the branches are more elongate, widely spaced, and cylindrical (Fig. 115). Less commonly, *P. (S.) stellata* is found to be encrusting, with irregular lobes, especially where wave action is heavy or in deeper water under ledges and on walls. The septa are well developed, continuous, and protrude well into the center, leaving a tiny circular hole. The upper septal surface is jagged and finely granulated (Fig. 116). The calices are shallow, unwalled depressions in some forms and are nearly flush with the surface in other forms (Fig. 115). The color of living colonies is pale brown or pale green with pale red patches. Previously, T.W. Vaughan (1907) described another species, *P. (S.) brighami,* which forms a continuous series with

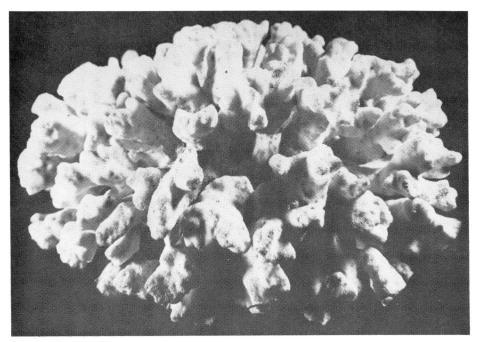

Figure 114.—SCLERACTINIA. *Psammocora (Stephanaria) stellata* Verrill.

Figure 115.—SCLERACTINIA. A colony of *Psammocora (Stephanaria) stellata* Verrill, originally described as *P. (S.) brighami* Vaughan.

Figure 116.—SCLERACTINIA. *Psammocora (Stephanaria) stellata* Verrill, close-up showing the arrangement of calices on the end of a branching projection.

typical specimens of *P. (S.) stellata*. Both are considered here to be synonyms, since differentiation between the species was based upon variable, intergrading growth form characters. *P. (S.) stellata* was originally described from Panama.

Psammocora verrilli Vaughan. This species is somewhat rare on Hawaiian reefs and is invariably found as encrustations which follow the shape of the surface over which the coral is growing (Fig. 117). The calices are situated in shallow depressions and are surrounded by crests that sometimes join together to form meandering ridges. These ridges are rounded and gentle compared to those of the other

Figure 117.—SCLERACTINIA. *Psammocora verrilli* Vaughan, type specimen.

Figure 118.—SCLERACTINIA. *Psammocora* sp. cf. *P. nierstraszi* Van der Horst; headlike specimen.

Hawaiian species of *Psammocora*. Although the calices have definite centers, they have indefinite boundaries and are variable in size (1.3 mm to 2.5 mm) and shape. The septa are thick with narrow interspaces. There are 2 or 3 complete cycles, none reaching the center of the calyx, which is a small hole occupied by a tiny columella. The upper edges of the septo-costae are minutely jagged. The coral has only been recorded at Hanauma Bay and Kaneohe Bay, Oahu, and Pukoo, Molokai, all on shallow reef flats far from shore. Colonies are pale brown or gray when alive.

Psammocora nierstraszi Van der Horst. An unidentified species of *Psammocora* (reported in Kaneohe Bay, Oahu; Makena, Maui; and Keauhou-Kona and Kapoho, Hawaii) closely matches the description of *P. nierstraszi*. However, there are some differences between the Hawaiian specimens and the species description, and positive identification of the coral must await careful review of similar species and firsthand comparisons of skeletal material. Hawaiian colonies are encrusting, sometimes sending up lobes. The lobes are composed of irregular knolls that rise with sharp edges, between which are smaller ridges and cones. The calices are irregularly scattered in groups or short rows between the ridges. The most prominent surface feature is the presence of many toothlike projections (dentacles) that cover all surfaces of the calices and coenosteum (Fig. 118). Because of the dentacles, the coral has a sandpaperlike texture. Living colonies take on an olive green coloration, sometimes crowned with vivid green patches when prominent lobes are present. Colonies range from 15 cm to nearly 1 m in diameter. The species has been reported in shallow tide pools and in deeper water (20 m depth) further offshore.

GLOSSARY (SCLERACTINIA)

Skeletal Characters

calyx (pl. **calices**): cuplike skeletal depression enclosed by walls and radiating partitions which house the anemonelike **polyp**.

coenosteum = **coenochyme, peritheca**: common surface mass of skeleton between the calices, if present. The various structures of the coenosteum are deposited by the **coenosarc**.

columella: a skeletal structure, often columnlike, in the center of the calyx formed by the fusion of the inner edges of the septa.

septum (pl. **septa**): thin vertical skeletal partitions fused to the inner surface of the walls of the calyx which project toward the center of the calyx. The septa divide the calyx into radial segments. Septa are arranged in regular **cycles**; the first cycle includes the largest septa, called the **primaries**, the next largest the **secondaries**, and so forth.

septo-costae: longitudinal ridgelike extensions of the septa outside the boundaries of the calices.

theca: walls of the calyx which enclose the aboral end of the polyp. The **epitheca** are elevated wall-like extensions of the theca that surround individual calices.

verrucae: small cylindrical projections arising from the skeleton of certain corals, particularly *Pocillopora*.

Colony Shape or Growth Form Characters

cespitose: bushy-shaped, with branches forming clumps or tufts.

columniform: formed of rounded columnlike projections.

encrusting: formed of continuous thin sheets everywhere adhering to the substratum. The form of the coral often assumes the substratum contours.

foliaceous = **explanate**: spreading or extending outward horizontally in a series of thin semicircular plates.

glomerate: lobed, massive, rounded or compact forms.

ramose: branched as the stem or root of a plant.

subspheroidal: composed of convex or knobby faces, or ball-like.

SURFACE CHARACTERS

ceroid: angular calices closely crowded and united directly to one another by thin walls.
foveolate: having small depressed calices.
meandroid: polyps (and calices) forming groups or series within common walls so as to form a valley. Sometimes used to describe the nature of the spaces between the branches of some ramose corals.
papillate: having rounded nipplelike projections **(papillae)** covering the coenosteum.
plocoid: with calices having distinct walls separated from each other by the coenosteum.
tuberculate: having minute rodlike projections **(tubercles)** on the coenosteum.

REFERENCES (SCLERACTINIA)

Agassiz, A.
  1889. The Coral Reefs of the Hawaiian Islands. *Bull. Mus. Comparative Zoology* 17(3): 121–170.
Alcock, A.
  1893. On Some Newly-Recorded Corals from the Indian Seas. *J. Asiatic Soc. Bengal* 62: 138–149. 3 pls.
Banner, A. H.
  1968. *A Fresh Water Kill on the Coral Reefs of Hawaii.* Univ. Hawaii Inst. Marine Biology Tech. Rep. 15.
Banner, A. H., and J. H. Bailey
  1970. *The Effects of Urban Pollution upon a Coral Reef System: A Preliminary Report.* Univ. Hawaii Inst. Marine Biology Tech. Rep. 25.
Barry, C. K.
  1965. Ecological Study of the Decapod Crustaceans Commensal with the Branching Coral *Pocillopora meandrina* var. *nobilis.* Master of Science Thesis, Univ. Hawaii.
Bernard, H. M.
  1897. The Genus Montipora: The Genus Anacropora. *British Mus. (Natural History) Catalogue Madreporarian Corals* (London). Vol. 3. 34 pls.
  1905. Porites of the Indo-Pacific Region. *British Mus. (Natural History) Catalogue Madreporarian Corals* (London). Vol. 5. 35 pls.
Bosch, H. F.
  1965. A Gastropod Parasite of Solitary Corals in Hawaii. *Pacific Science* 19(2): 267–268.
  1967. Growth Rate of *Fungia scutaria* in Kaneohe Bay, Oahu. Master of Science Thesis, Univ. Hawaii.
Boschma, H.
  1923. The Madreporaria of the Siboga Expedition: Pt. 4, *Fungia patella. Siboga Expedition Monogr.* 16d: 129–148. 2 pls.
  1926. Madreporaria I: Fungiidae. *Videnskabelige Meddelelser Dansk Naturhistorisk Forening Khobenhavn* 79: 185–259.
  1954. On Some Specimens of the Coral *Montipora verrucosa* (Lamarck) from the Hawaiian Islands, Formerly Described as Separate Species. *Proc. Koninklijke Nederlandse Akad. Wetenschappen,* Ser. C, *Biological and Medical Sci.* 57(2): 151–158.
Branham, J. M., S. A. Reed, J. H. Bailey, and J. Caperon
  1971. Coral-Eating Sea Stars *Acanthaster* in Hawaii. *Science* 172: 1155–1157.
Brook, G.
  1893. The Genus Madrepora. *British Mus. (Natural History) Catalogue Madreporarian Corals* (London). Vol. 1. 35 pls.
Bryan, W. A.
  1915. *Natural History of Hawaii.* Honolulu: Hawaiian Gazette.
Chave, K. E.
  1967. Recent Carbonate Sediments: An Unconventional View. *J. Geological Education* 15(5): 200–204.
Dana, J. D.
  1846. *Zoophytes.* U.S. Exploring Expedition (Washington). Vol. 7. 61 pls.
  1872. *Corals and Coral Islands.* New York: Dodd, Mead.

Dana, T. F.
   1971. On the Reef Corals of the World's Most Northern Atoll (Kure: Hawaiian Archipelago). *Pacific Science* 25(1): 80–87.

Döderlein, L.
   1901. Die Korallen-Gattung Fungia. *Zoologischer Anzeiger* 24: 353–360.
   1902. Die Korallen-Gattung Fungia. *Abhandlungen Senckenbergischen Naturforschenden Gesellschaft* 27(1): 1–162.

Dollar, S. J.
   1975. Zonation of Reef Corals off the Kona Coast of Hawaii. Master of Science Thesis, Univ. Hawaii.

Edmondson, C. H.
   1924. Notes on the Rate of Growth of Coral in Hawaii. *Proc. Second Pan-Pacific Science Congress* (Australia) 2: 1553–1555.
   1928. *The Ecology of a Hawaiian Coral Reef.* B. P. Bishop Mus. Bull. 45. Honolulu.
   1929. *Growth of Hawaiian Corals.* B. P. Bishop Mus. Bull. 58. Honolulu.
   1933a. *Reef and Shore Fauna of Hawaii.* B. P. Bishop Mus. Spec. Publ. 22. Honolulu.
   1933b. *Cryptochirus* of the Central Pacific. *B. P. Bishop Mus. Occasional Pap.* 10(5): 3–23.
   1946a. Behavior of Coral Planulae under Altered Saline and Thermal Conditions. *B. P. Bishop Mus. Occasional Pap.* 18(19): 283–304.
   1946b. *Reef and Shore Fauna of Hawaii.* B. P. Bishop Mus. Spec. Publ. 22 (Rev. ed.). Honolulu.

Fowler, G. H.
   1888. The Anatomy of the Madreporaria. *Quart. J. Microscopical Science* 28: 413–430. 33 pls.

Franzisket, L.
   1970. The Atrophy of Hermatypic Corals and Their Subsequent Regeneration in Light. *Internationale Revue gesamten Hydrobiologie* 55(1): 1–12.

Galtsoff, P. S.
   1933. *Pearl and Hermes Reef, Hawaii: Hydrographical and Biological Observations.* B. P. Bishop Mus. Bull. 107. Honolulu.

Gordon, M. S., and H. M. Kelly
   1962. Primary Productivity of a Hawaiian Coral Reef: A Critique of Flow Respirometry in Turbulent Water. *Ecology* 43(3): 473–480.

Grigg, R. W., and J. E. Maragos
   1974. Recolonization of Hermatypic Corals on Submerged Lava Flows in Hawaii. *Ecology* 55(2): 387–395.

Harrigan, J. F.
   1972. The Planula Larva of *Pocillopora damicornis:* Lunar Periodicity of Swarming and Substratum Selection Behavior. Doctor of Philosophy Dissertation, Univ. Hawaii.

Hiatt, R. W.
   1954. Hawaiian Marine Invertebrates: A Guide to Their Identification. Unpubl. ms., Sinclair Library, Univ. Hawaii.

Hobson, E. S.
   1974. Feeding Relationships of Teleostean Fishes on Coral Reefs in Kona, Hawaii. *Fishery Bull.* 72(4): 915–1031.

Hoffmeister, J. E.
   1925. *Some Corals from American Samoa and the Fiji Islands.* Pap. Dept. Marine Biology, Carnegie Inst. Washington, Vol. 22. Carnegie Inst. Washington Publ. 343. Washington, D.C.

Johannes, R. E., S. L. Coles, and N. T. Kuenzel
   1970. The Role of Zooplankton in the Nutrition of Some Scleractinian Corals. *Limnology and Oceanography* 15(4): 579–586.

Jokiel, P. L., and H. P. Cowdin
   1976. Hydromechanical Adaptation in the Solitary Free-Living Coral *Fungia scutaria. Nature* 262: 212–213.

Jokiel, P. L., and S. J. Townsley
   1974. Biology of the Polyclad *Prostiostomum* ns.: A New Coral Parasite from Hawaii. *Pacific Science* 28(4): 361–373.

Kohn, A. J., and P. Helfrich
   1957. Primary Organic Productivity of a Hawaiian Coral Reef. *Limnology and Oceanography* 2: 241–251.

Lamarck, J. B. P.
    1801. *Systeme des Animaux sans Vertebres.* Paris.
    1816. *Histoire Naturelle des Animaux sans Vertebres II.* Paris.
Lenhoff, H., L. Muscatine, and L. V. Davis
    1967. Coelenterate Biology: Experimental Research. *Science* 160: 1141–1145.
Lesson, R. P.
    1831. *Illustrations de Zoologie.* Paris. 60 pls.
MacCaughey, V.
    1918. A Survey of the Hawaiian Coral Reefs. *American Naturalist* 52: 409–438.
Mackaye, A. C.
    1915. Corals of Kaneohe Bay. *Hawaiian Almanac and Annual for 1916,* pp. 135–139.
Maragos, J. E.
    1972. A Study of the Ecology of Hawaiian Reef Corals. Doctor of Philosophy Dissertation, Univ. Hawaii.
    1974. Reef Corals of Fanning Island. *Pacific Science* 28(3): 247–255.
Maragos, J. E., and P. L. Jokiel
    In Press. Reef Corals of Canton Atoll, Phoenix Islands: Notes on Coral Zoogeography. *Atoll Res. Bull.* Washington, D.C.: Smithsonian Inst.
Menard, H. W., E. C. Allison, and J. W. Durham
    1962. A Drowned Miocene Terrace in the Hawaiian Islands. *Science* 138: 896–897.
Milne-Edwards, H., and J. Haime
    1848–1850. Recherches sur les Polypes. *Annales Sciences Naturelles* (Paris). Vols. 9–13.
    1857–1860. *Histoire Naturelle des Coralliaires.* 3 vols. Paris.
Neighbor Island Consultants, Inc.
    1977. *Environmental Surveys of Deep Ocean Dredged Spoil Disposal Sites in Hawaii.* U.S. Army Engineer Div., Pacific Ocean, Corps of Engineers.
Oceanic Foundation
    1975. A Three Year Environmental Study of Honokohau Harbor, Hawaii. Honolulu: U.S. Army Engineer Div., Pacific Ocean, Corps of Engineers.
Pollock, J. B.
    1928. *Fringing and Fossil Reefs of Oahu.* B. P. Bishop Mus. Bull. 55. Honolulu.
Powers, D. A.
    1970. A Numerical Taxonomic Study of Hawaiian Reef Corals. *Pacific Science* 24(2): 180–186.
Powers, D. A., and F. James Rohlf
    1972. A Numerical Taxonomic Study of Caribbean and Hawaiian Reef Corals. *Systematic Zoology* 21(1): 53–64.
Quelch, J. J.
    1886. Report on the Reef Corals. *Report of the Scientific Results of the Voyage of H.M.S. Challenger, Zoology* Vol. 16, pt. 3. 12 pls. London.
Reed, A. S.
    1971. Some Common Coelenterates in Kaneohe Bay, Oahu, Hawaii. In H. M. Lenhoff, L. Muscatine, and L. V. Davis (eds.), *Experimental Coelenterate Biology,* pp. 37–51. Honolulu: Univ. Hawaii Press.
Sorokin, Y. I.
    1973. Microbiological Aspects of the Productivity of Coral Reefs. In O. A. Jones and R. Endean (eds.), *Biology and Geology of Coral Reefs,* Vol. 2, Biology 1, pp. 17–45. New York: Academic Press.
Squires, D. F.
    1965. Neoplasia in Coral? *Science* 148(3669): 503–505.
Stearns, H. T.
    1966. *Geology of the State of Hawaii.* Palo Alto: Pacific Books.
Stephens, G. C.
    1960. Uptake of Glucose from Solution by the Solitary Coral *Fungia. Science* 131(3412): 1532.
    1962. Uptake of Organic Material by Aquatic Invertebrates: I, Uptake of Glucose by the Solitary Coral, *Fungia scutaria. Biological Bull.* 123(3): 648–659.
Studer, T.
    1901. Madreporarier von Samoa, den Sanwich-Inseln und Laysan. *Zoologische Jahrbücher Systematik* 14(5): 388–428.

Van der Horst, C. J.
    1922. The Madreporaria of the Siboga Expedition: Pt. 2, Madreporaria Fungida. *Siboga Expedition Monogr.* 16b: 53–98. 6 pls.

Vaughan, R. A.
    1973. Aspects of the Ecology of *Alpheus deuteropus* (Crustacea, Decapoda), a Boring Shrimp. Master of Science Research Rep. Univ. Hawaii, Dept. Zoology.

Vaughan, T. W.
    1905. A Critical Review of the Literature of the Simple Genera of the Madreporidae, Fungida, With a Tentative Classification. *Proc. U.S. National Mus.* 28: 371–424.
    1907. *Recent Madreporaria of the Hawaiian Islands and Laysan.* U.S. National Mus. Bull. 59. Washington, D.C.
    1910. Summary of the Results Obtained from a Study of the Recent Madreporaria of the Hawaiian Islands and Laysan. *Proc. Seventh International Zoological Congress.* (BPBM QL Pam. 227.)

Vaughan, T. W., and J. Wells
    1943. *Revision of the Suborders, Families and Genera of the Scleractinia.* Geological Soc. America Spec. Pap. 44.

Verrill, A. E.
    1864. List of Polyps and Corals Sent by the Museum of Comparative Zoology to the Other Institutions in Exchange, with Annotation. *Bull. Mus. Comparative Zoology Harvard* 1(3): 29–60.
    1865. Polyps and Corals of the Pacific Exploring Expedition. *Proc. Essex Inst.* 4: 182.
    1885. Reports on the Results of Dredging by the United States Coast Guard Survey Steamer "Blake" XXI. Report on the Anthozoa and on Some Additional Species Dredged by the "Blake" in 1878–79, and by the U.S. Fish Commission Steamer "Fish Hawk" in 1880–82. *Bull Mus. Comparative Zoology Harvard* 11: 1–72. 8 pls.
    1902. Notes on the Genus *Acropora* (Madrepora Lmk.) with New Descriptions and Figures of Types, and of Several New Species. *Trans. Connecticut Acad. Arts Sciences* 11: 207–266.

Wells, J.
    1954. *Recent Corals of the Marshall Islands.* U.S. Geological Survey Prof. Pap. 260-I.

Yonge, C. M.
    1940. The Biology of Reef Building Corals. *Great Barrier Reef Expedition 1928–1929* 1(13): 353–391.

ORDER ANTIPATHARIA
BLACK CORALS
RICHARD W. GRIGG and DENNIS OPRESKO

University of Hawaii and
Institute of Marine and Atmospheric Sciences, Florida

ANTIPATHARIANS, commonly known as black corals, occur in all oceans, ranging from just below the low tide mark to depths of thousands of meters. The majority of the 150 recognized species are found in tropical seas below the euphotic zone (>100 m). However, a few species occur in water shallower than 20 meters, but these generally are found where there is very little light, such as in caves and on the undersurfaces of ledges. The ecological aspects of the vertical distribution of one species of antipatharian, *Antipathes grandis*, have been studied by Grigg (1963, 1965), but because of the remoteness of the habitats that most black corals occupy, very little is known about their biology and life history (Pax, 1940). The black corals, because of their sessile (fixed) nature, attract a number of symbiotic organisms including worms, crustaceans, mollusks, and fishes (Grigg, 1963; Davis and Cohen, 1968). The antipatharians offer shelter, as well as a substratum for residence or attachment of these animals.

Antipatharians secrete an axial skeleton which is usually dendritic (treelike) in form; however, some species are encrusting, while others form long, unbranched, whiplike colonies. The skeleton of black coral is actually brown or reddish brown in color, only appearing black in some species because it is generally opaque and quite dense. The horny skeleton is similar to gorgonin (Roche and Eysseric-Lafon, 1951), the protein material that comprises the skeletal axis of holaxonian gorgonians. Skeletons of antipatharians and gorgonians, however, can readily be distinguished by the presence of spines on the surface of the former (Fig. 1). The black corals, in fact, are sometimes referred to as *dörnchenkorallen* (little thorn corals).

A single antipatharian colony can possess thousands of tiny polyps which form a "living bark" or cortex around the skeleton. In all but one of the described species, the polyps have six unbranched, nonretractile tentacles surrounding the mouth. The polyps are gelatinous and often a white, red, or orange color. They differ from the polyps of gorgonians in that they lack calcareous spicules and pinnate tentacles. Many species are armed with nematocysts, which may have an adverse effect on divers collecting and handling live colonies.

The skeletons of some of the larger species of black coral (colonies of *Antipathes grandis* commonly reach 2 m in height) have long been used for jewelry. Historical aspects of the black coral jewelry industry have been described by Hickson (1924). One of the earliest fisheries for black coral existed in the Red Sea several thousand years ago. The corals were easily marketed in the Far East, where

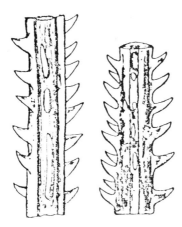

Figure 1.—ANTIPATHARIA. *Antipathes grandis.* Sections of larger branches showing spines (from Verrill, 1928).

they were used as scepters, divining rods, and as amulets to ward off evil and injury. According to a Greek myth, black coral was originally given to Ulysses as a protective charm against Circe. The natives of North Africa depended on polished objects of black coral to neutralize the magic of the dreaded "evil eye." It was because of such practices that the name *Antipathes* (*Anti,* "against"; *pathos,* "disease or suffering") was derived.

The Hawaiian fishery for black coral began in 1958. Stewart (1962a) gives an early account of the first discovery of this precious coral in local waters, and in a later paper (1962b), describes some of the methods used in working black coral into jewelry. Recently, Poh (1971) reported on the economics and market potential of the industry in Hawaii.

Although there have been a number of major taxonomic monographs written on the Antipatharia—including those of Brook (1889), Schultze (1896), Cooper (1909), van Pesch (1914), and Pax (1918)—the systematics of the order is in great need of revision. According to Cooper (1909), from a taxonomic standpoint the number of mesenteries, presence of a single or branching axis, unity or partial division of the polyps, and other morphological features enable specimens to be placed into recognizable genera such as *Cirrhipathes, Stichopathes,* and *Antipathes.* However, the characters that serve to distinguish species in these groups are poorly defined and many of the species are so badly described that misidentifications are inevitable. These limitations have been taken into consideration in dealing with the species presented here. In some cases only tentative identifications can be made until more detailed taxonomic studies are completed.

Prior to this work only three species of antipatharians had been reported from Hawaiian waters. Two of these species—*Antipathes grandis* and *Antipathes irregularis*—were originally described by Verrill (1928). The first species is valid, and it is used extensively in the black coral jewelry industry. The second species, which was taken from the stomach of a shark, has been re-examined, and the lack of skeletal spination indicates that this is a gorgonian rather than an antipatharian. The third species closely resembles *A. grandis* and was first collected off Lahaina, Maui, in 1958. It was later identified by Bayer (1961) as *Antipathes dichotoma* Pallas. The recent Sango Expedition of the University of Hawaii Sea Grant program has resulted in the discovery of twelve additional species of antipatharians in Hawaiian waters. Six of these species belong to the genus *Antipathes;* two, to the genus *Cirrhipathes;* and one each, to the genera *Stichopathes, Leiopathes, Schizopathes,* and *Parantipathes.* Locations and depths of all species from Hawaii are given in Table 1.

TABLE 1
Species of Antipatharia from Hawaii

Species	Location	Depth(m)
Antipathes dichotoma Pallas 1766	Hawaii to Niihau	30–85**
Antipathes grandis Verrill 1928	Hawaii to Kauai	50–85**
Antipathes intermedia (Brook 1889)	Oahu, 157°32.8′W–21°18.0′N	347–366*
Antipathes punctata (Roule 1905)	Oahu, 157°32.8′W–21°18.2′N	300*
	Oahu, 157°32.8′W–21°18.5′N	395–425*
	Oahu, 158°23.2′W–21°35.1′N	337–392*
	Oahu, 158°24.3′W–21°36.2′N	395–445*
Antipathes subpinnata Ellis and Solander 1786	Oahu, 158°24.6′W–21°35.8′N	455–460*
Antipathes ulex Ellis and Solander 1786	Hawaii to Oahu	30–50**
	Brooks Banks, 166°40.1′W–23°59.0′N	216–330*
Antipathes undulata van Pesch 1914	Oahu, 158°04.9′W–21°15.1′N	234–490*
	Oahu, off Barbers Point	183–219*
	Oahu, off Kaneohe Bay	110–274*
	Brooks Banks, 166°43.5′W–24°00.3′N	384–432*
Antipathes sp.	Brooks Banks, 166°41.9′W–23°56.0′N	373–430*
	French Frigate Shoals, 165°22.1′W–23°55.0′N	353–381*
Cirrhipathes anguina Dana 1846	Hawaii to Kauai	25–40**
Cirrhipathes spiralis (Linnaeus 1758)	Kauai, 159°47.5′W–22°11.4′N	397–410*
	Kauai, 159°20.9′W–21°53.4′N	408–454*
	Oahu, 157°56.4′W–21°46.8′N	326–436*
Leiopathes glaberrima (Esper 1788)	Oahu, 157°42.6′W–21°13.8′N	378+
	Oahu, 157°32.0′W–21°18.3′N	400+
	Kauai, 159°44.8′W–21°56.7′N	295–432*
Parantipathes sp.	Oahu, 158°24.0′W–21°37.2′N	326–447*
Schizopathes conferta Brook 1889	Oahu, 158°25.0′W–21°36.0′N	380+
Stichopathes echinulata Brook 1889	Lanai, 156°58.7′W–20°36.6′N	305–565*

* = depth range of dredge haul
** = collected using SCUBA
+ = collected by the submersible, *Star II*

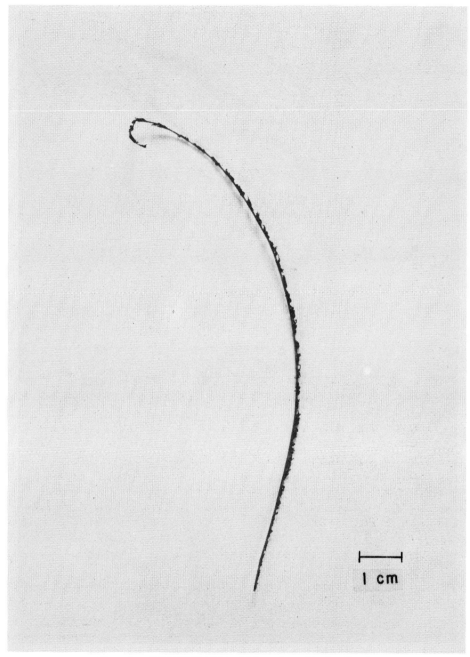

Figure 2.—ANTIPATHARIA. *Stichopathes echinulata* Brook.

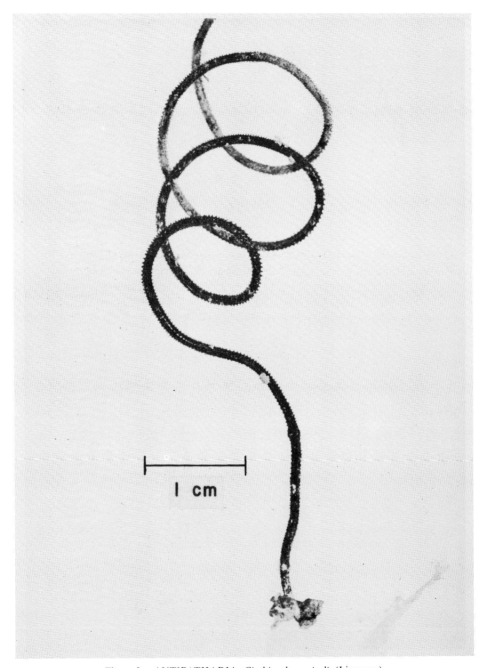

Figure 3.—ANTIPATHARIA. *Cirrhipathes spiralis* (Linnaeus).

Figure 4.—ANTIPATHARIA. *Cirrhipathes anguina* Dana. South Point, Island of Hawaii, depth 150 ft. 1868 lava flow (photo by R. Grigg).

KEY TO ANTIPATHARIAN SPECIES KNOWN FROM HAWAIIAN WATERS

1 Colony unbranched (very rarely with branches arising from damaged parts of axis) 2
  Colony branched .... [*Antipathes, Parantipathes, Leiopathes, Schizopathes*] 4

2(1) Polyps in more than one longitudinal row along length of axis ... [*Cirrhipathes*] 3
  Polyps in single row ... [*Stichopathes*][1] ....
  Spines minute, triangular (in 10 to 15 longitudinal rows) near tip of colony, becoming longer on thicker portions of axis (Fig. 2) .... *Stichopathes echinulata* Brook[2]

---

[1] In 1910 van Pesch synonymized *Stichopathes* with *Cirrhipathes*, but he separated the two groups subgenerically. However, more recent workers such as Pax (1918) have continued to treat *Stichopathes* as a valid genus. The taxonomic features separating the two groups are, in some cases, not very distinct. The tendency for the polyps to occur on more than one side of the axis is largely dependent on polyp density, which is a continuously variable character. Species exhibiting either extreme of this character can be placed rather easily in *Cirrhipathes* or *Stichopathes*, but intermediates are much more difficult to classify. Until a careful taxonomic revision is undertaken, the generic or subgeneric status of these two groups will remain unclear. Preliminary evidence suggests, however, that some species now placed in *Cirrhipathes* may even prove to be more closely related to species of *Antipathes* than to other species of *Cirrhipathes* or *Stichopathes*.

[2] In the size and shape of its polyps and spines, this species appears to be closely related to *Stichopathes robusta* Gravier; however, in *S. echinulata* the spines have a much more regular arrangement on the axis. On the lower parts of the corallum, where the spines tend to be larger and farther apart, the species resembles, to some degree, *Stichopathes paucispina* Brook.

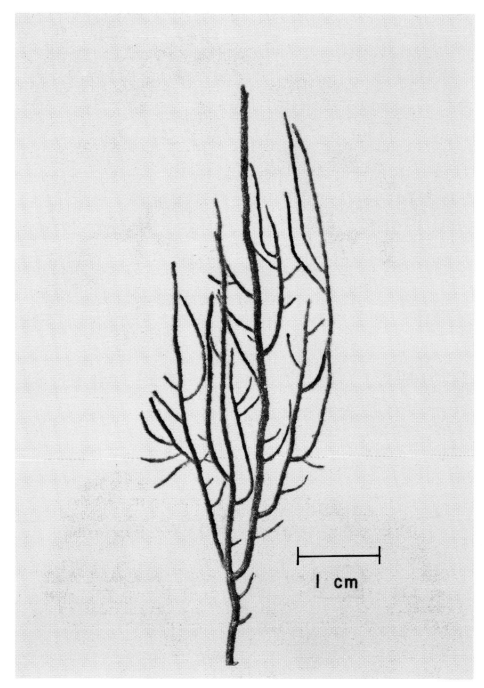

Figure 5.—ANTIPATHARIA. *Antipathes undulata* van Pesch.

Figure 6.—ANTIPATHARIA. *Antipathes subpinnata* Ellis and Solander.

3(2) Colony often in tight spirals 1 cm to 2 cm in diameter near tip; spines triangular, smooth or slightly papillose and longer on polyp side of axis which is usually on the convex side of the spirals
(Fig. 3) .... *Cirrhipathes spiralis* (Linnaeus)
Colony straight or weakly spiral, often with nodelike swellings occurring at irregular intervals; spines conical, inclined upward and longer on polyp side of axis (Fig. 4)............... *Cirrhipathes anguina* Dana

4(1) Smallest branches (pinnules) not arranged regularly, but may be confined to a single side or to opposite sides of larger branches 5
Smallest branches (pinnules) of almost equal length and tending to be arranged regularly and symmetrically on larger branches 10

5(4) Smallest branches arranged bilaterally on opposite sides of larger branches but of unequal length and spaced at varying intervals 6
Smallest branches placed irregularly on all sides of larger branches 7

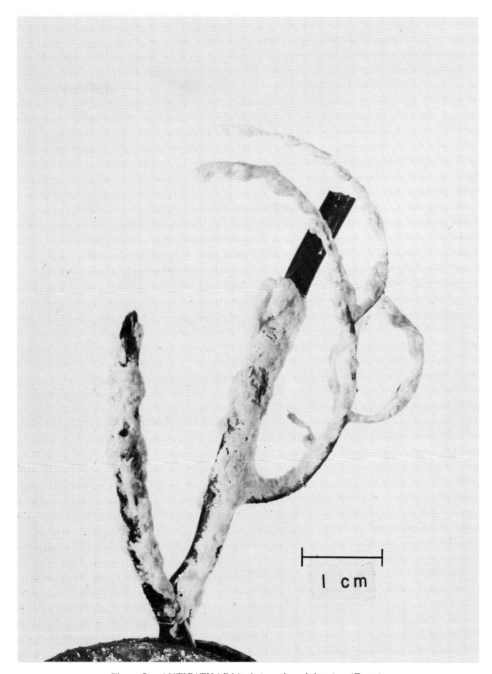

Figure 7.—ANTIPATHARIA. *Leiopathes glaberrima* (Esper).

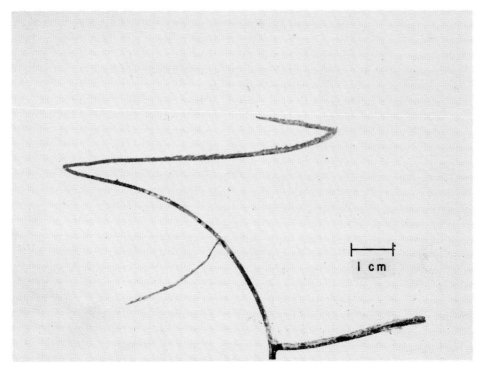

Figure 8.—ANTIPATHARIA. *Antipathes* sp.

6(5) Colony flabellate, smallest branches short and curved upward; spines variable in size, longest around circumference of polyps (Fig. 5) .... *Antipathes undulata* van Pesch
Colony not branched in a single plane, smallest branches usually 2 mm to 3 mm apart, mostly 1 cm long and directed upward; spines conical and acute, becoming longer on thicker branches (Fig. 6) .... *Antipathes subpinnata* Ellis and Solander[3] (*sensu* Brook)

7(5) Branches appearing crooked and bent, often arising at right angles from larger branches; spines extremely small, triangular, often with rounded apex, absent from larger branches; polyps with 12 septa (Fig. 7) .... *Leiopathes glaberrima* (Esper)
Branches not appearing crooked and bent, often very long; spines triangular or conical and usually slightly larger on polyp side of axis; polyps with 10 septa 8

---

[3]The specimens identified here as *Antipathes subpinnata* agree with the descriptions of this species as given by Brook (1889) and later workers; however, they differ from Ellis and Solander's original description in that the branching pattern is more irregular.

Figure 9.—ANTIPATHARIA. *Antipathes dichotoma* Pallas.

Figure 10.—ANTIPATHARIA. Off Lahaina, Maui, in the Auau Channel, *Antipathes dichotoma* occurs in abundance along steep drop-offs at depths between 120 and 350 feet. Photo taken at 160 feet.

8(7) Branching sparse, consisting of only a few very long branches
 (Fig. 8) .... *Antipathes* sp.[4]
 Branching dense 9
9(8) Smallest branches 0.6 mm to 0.7 mm in diameter, stiff and directed
 vertically; spines faintly papillose or with small knobs near apex;
 7 to 8 polyps per cm (Figs. 9, 10) ...... *Antipathes dichotoma* Pallas[5]
 Smallest branches 0.3 mm in diameter, thin and flexible; spines smooth
 or faintly papillose; 11 to 12 polyps per cm (Fig. 11)
 .... *Antipathes grandis* Verrill
10(4) Pinnules in two lateral rows along length of main axis or branches 11
 Pinnules in four rows along length of branches 13
11(10) Pinnules opposite, grouped together in pairs generally 6 mm to 9 mm
 apart (Fig. 12) .... *Antipathes punctata* (Roule)
 Pinnules arranged alternately 12

---

[4]This species cannot be identified with certainty at this time, but it appears to be related to *Antipathes galapagensis* Deichmann 1941.

[5]The axial spines of this species are very similar to those of *Antipathes lentipinna* Brook, one of 19 species synonymized with *A. dichotoma* by van Pesch (1914). Additional studies may show that *A. lentipinna* is a distinct species. Our specimens differ from those of Brook in that the branches are more vertically inclined.

Figure 11.—ANTIPATHARIA. *Antipathes grandis* Verrill.

12(11) Corallum monopodial, consisting only of a long main axis (stem) and two lateral rows of simple pinnules; pinnules equidistant and nearly equal in length (Figs. 13, 14); subpinnules absent
.... *Schizopathes conferta* Brook[6]
Corallum branched, pinnules arising from stem and branches, but not always evenly spaced and not exactly equal in length; subpinnules usually present on the anterior side and near base of primary pinnules (Fig. 15) .... *Antipathes ulex* Ellis and Solander[7]
13(10) Pinnules in the two anterior rows often with one or two subpinnules arising near point of insertion; polyps very elongate
(Fig. 17) .... *Parantipathes* sp.
Pinnules in the two anterior rows usually unbranched; polyps short
(Fig. 16) .... *Antipathes intermedia* (Brook)

---

[6]The specimens identified here as *Schizopathes conferta* Brook are in close agreement with Brook's (1889) description with the following qualification. Unlike the type of *S. conferta,* the Hawaiian specimens all have a basal plate. However, because Brook's specimen is only part of a corallum, broken off above the base, there is no way to determine if it too had an adherent basal plate. Brook placed this species in *Schizopathes* in view of the fact that it has very crowded polyps like all other known species of that genus, but all species of *Schizopathes* that have been described from complete specimens form unattached colonies. They produce a bulbous expansion near the lower end of the stem and use this to anchor themselves in the soft bottom mud, very much like a pennatulid. If our identification is correct, it would suggest that either the definition of *Schizopathes* should be enlarged to include both attached and unattached forms, or that a new genus should be established for attached forms having crowded polyps.

[7]A closely related, if not identical species, is *Antipathes japonica* Brook.

GLOSSARY (ANTIPATHARIA)

apex: tip.
axis: (axes, pl.): central supporting skeletal structure of colony.
basal plate: flattened adherent base of a colony.
corallum: skeleton.
flabellate: fan-shaped.
papillose: covered by small protuberances or papillae.
pinnules: smallest branches, usually unbranched.
subpinnules: small secondary branches of pinnules in some forms.

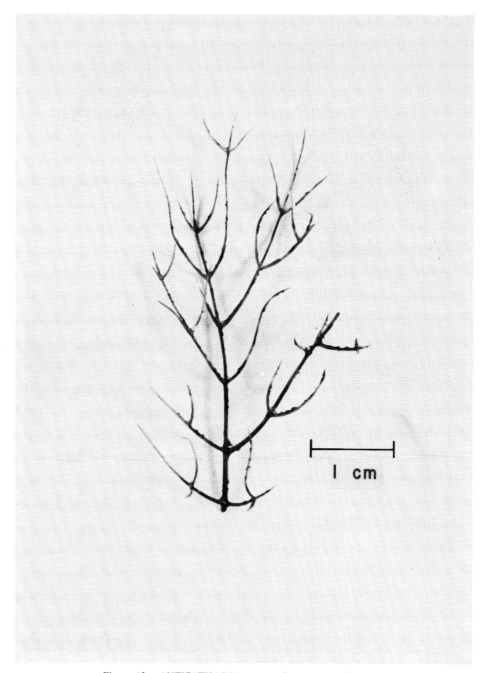

Figure 12.—ANTIPATHARIA. *Antipathes punctata* (Roule).

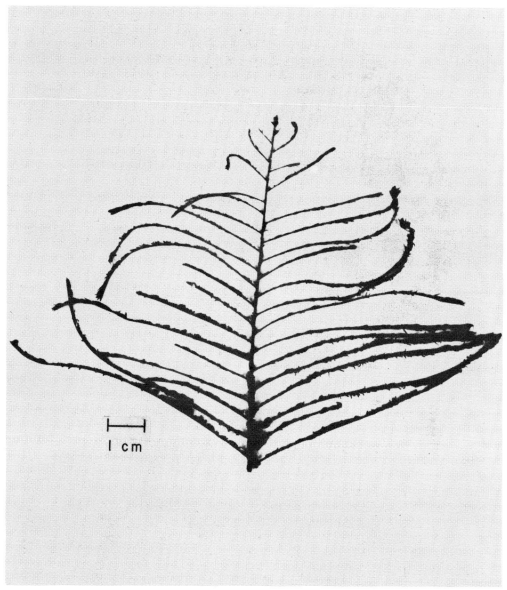

Figure 13.—ANTIPATHARIA. *Schizopathes conferta* Brook.

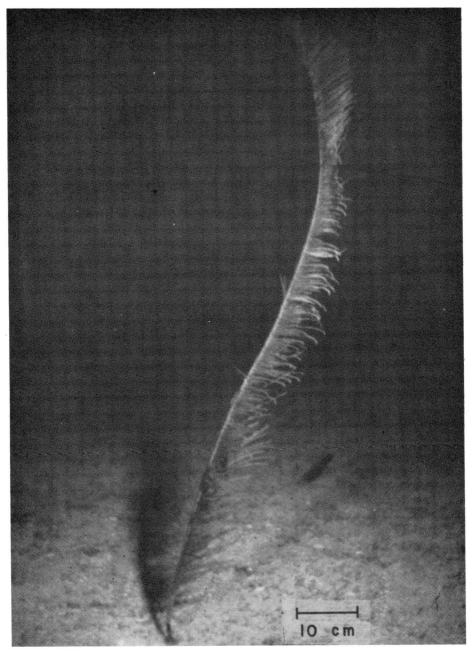

Figure 14.—ANTIPATHARIA. *Schizopathes conferta.* Photo taken from submersible *Star II*, at 1,125 feet depth off Makapuu, Oahu.

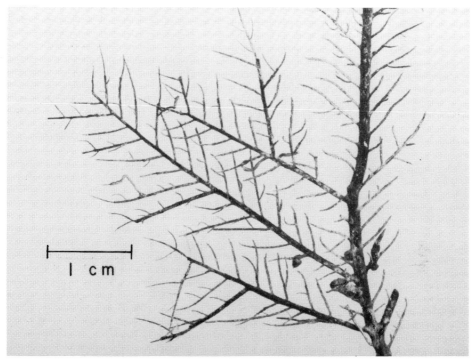

Figure 15.—ANTIPATHARIA. *Antipathes ulex* Ellis and Solander.

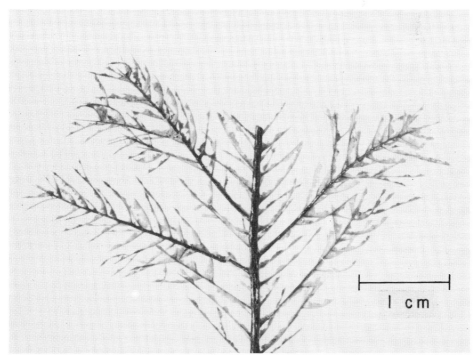

Figure 16.—ANTIPATHARIA. *Antipathes intermedia* (Brook).

Figure 17.—ANTIPATHARIA. *Parantipathes* sp.

REFERENCES (ANTIPATHARIA)

Bayer, F. M.
    1961. Summary of Black Coral Investigations to Date (Letter). *Hawaiian Shell News* 9(12): 8.

Brook, G.
    1889. Report on the Antipatharia. *Report of the Scientific Results of the Voyage of H.M.S. Challenger, Zoology* Vol. 32, Pt. 80. 15 pls.

Cooper, C. F.
    1909. Antipatharia. Reports of the Percy Sladen Trust Expedition to the Indian Ocean in 1905. *Trans. Linnean Soc. London,* Ser. 2A. 12(17): 301–321.

Dantan, J. L.
    1921. Recherches sur les Antipathaires. *Arch. Analytical Microscopy* 17: 137–245.

Davis, W. P., and D. M. Cohen
    1968. A Gobiid Fish and a Palaemonid Shrimp Living on an Antipatharian Sea Whip in the Tropical Pacific. *Bull. Marine Science* 18(4): 749–761. 6 figs.

Deichmann, E.
    1941. Coelenterates Collected on the Presidential Cruise of 1938. *Smithsonian Miscellaneous Collections* 99(10): 1–17. 1 pl.

Grigg, R. W.
    1963. A Contribution to the Biology and Ecology of the Black Coral, *Antipathes grandis* Verrill in Hawaii. Master's Thesis, Univ. Hawaii.
    1965. Ecological Studies of Black Coral in Hawaii. *Pacific Science* 19(2): 244–260.

Hickson, S. J.
    1924. *An Introduction to the Study of Recent Corals.* London: Manchester Univ. Press.

Pallas, D. S.
    1766. *Elenchus Zoophytorum.* The Hague.

Pax, F.
    1918. Die Antipatharien. *Zoologische Jahrb.* 41: 419–478. 85 figs.
    1940. Anthozoa. Ordnung: Antipatharia oder Dörnchenkorallen. In H. G. Bronns, *Klassen und Ordnungen das Tierreichs,* pp. 199–336. Vol. 2, Pt. 2, Book 3. Leipzig: Akademische Verlag.

Poh, K. K.
    1971. *Economics and Market Potential of the Precious Coral Industry in Hawaii.* R. W. Grigg (ed.). UNIHI-Sea Grant-AR-71-03. 22 pp.

Roche, J., and M. Eysseric-Lafon
    1951. Biochimie Comparee des Scleroproteines Iodees des Anthozoaires. *Bull. Soc. Chimie Biologique* 33(1): 1437–1447.

Schultze, L. S.
    1896. Beiträg zur Systematik der Antipatharien. *Abh. Senckenbergischen Naturforschenden Gesellschaft* 23(1): 1–39.

Stewart, J.
    1962a. The Hawaiian Black Coral Story: Part I. *Lapidary J.* July, p. 388.
    1962b. The Hawaiian Black Coral Story: Part II. *Lapidary J.* August, p. 490.

Tischbierek, H.
    1936. Die Nesselkapseln der Antipatharien. Thesis, Univ. Breslau. (Abstract in *Zoologische Berichte* 43(504).)

Van Pesch, A. J.
    1910. *Beiträge zur kenntnis der Gattung Cirrhipathes.* Leiden. 5 pls.
    1914. The Antipatharia of the Siboga Expedition. *Siboga Expedition Monogr.* 17. 8 pls.

Verrill, A. E.
    1928. *Hawaiian Shallow Water Anthozoa.* B. P. Bishop Mus. Bull 49. Honolulu.

CTENOPHORA

DENNIS M. DEVANEY
B. P. Bishop Museum

ANIMALS in the phylum Ctenophora are commonly referred to as comb jellies. The majority are planktonic although some members of the order Platyctenea occur as symbionts on algae, coelenterates, and echinoderms. Generally speaking, the ctenophores are characterized by biradial symmetry, eight meridional rows of ciliary or comb plates (called ctenes), and a pair of tentacles—often retractable into sheaths or pits—that are provided with adhesive cells called colloblasts. Most forms have a delicate gelatinous consistency and are inconspicuous because of their transparency, although some are capable of bioluminescence or have brilliantly pigmented areas.

Representatives of the five recognized orders have been observed in shallow Hawaiian waters (Matthews, 1954).

KEY TO THE ORDERS OF HAWAIIAN CTENOPHORES

1. Species having tentacles, usually throughout life 2
   Species lacking tentacles .... Order BEROIDA
2(1). Planktonic forms... 3
   Benthic, symbiotic forms (genus *Coeloplana*) ...... Order PLATYCTENEA
3(2). Globular forms with well-developed, retractile
       tentacles .... Order CYDIPPIDA
   Moderately to extremely compressed forms with tentacles more or less
       reduced in adult stage.. 4
4(3). Ovate and moderately compressed forms, having two large oral lobes and
       slender, flaplike ciliated auricles surrounding the
       mouth .... Order LOBATA
   Extremely compressed and laterally expanded, producing long ribbonlike
       forms; oral lobes and auricles absent .... Order CESTIDA

Order BEROIDA

Beroe forskali Milne-Edwards 1841. Among the ctenophores reported from the 1902 *Albatross* expedition in Hawaiian waters, Mayer (1906, p. 1134) mentioned "a *Beroe* apparently identical with *B. australis* Agassiz and Mayer [known] from the Ellice and Fiji islands," without giving further details. Bigelow (1912) considered *Beroe australis* synonymous with *B. forskali*, a wide-ranging Pacific species.

The specimen described by Agassiz and Mayer (1899) as *B. australis* was 40 mm long (Fig. 1a) with the body compressed laterally, one side about three times as

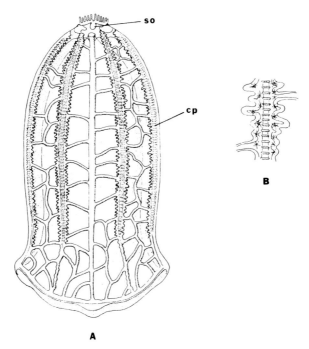

Figure 1.—CTENOPHORA. *Beroe forskali* Milne-Edwards. *a,* View of broad side; *cp,* ciliated plate; *so,* sense organ; *b,* portion of ciliated plate with stellate pigment cells (redrawn from Agassiz and Mayer, 1899, by David Kemble, Bishop Museum).

broad as the other. The aboral sense organ is surrounded by a figure-8-shaped row of branched papillae. The eight rows of ciliated plates are bordered by masses of stellate pigment cells (Fig. 1*b*). This species has been reported from surface waters (Agassiz and Mayer, 1899; 1902) and plankton hauls, at depths of 300 fathoms to 0 fathoms (Bigelow, 1912).

Order PLATYCTENEA

The platyctenids are a highly modified group of ctenophores in which the body is usually compressed in the oral-aboral plane. The result is a generally flattened form having two reduced, sheathed tentacles capable of great elongation (Fig. 2). Combs may be absent as they are in *Coeloplana,* a benthic genus of symbiotic forms represented in Hawaii by several species.

Coeloplana (?) *duboscqui* Dawydoff 1930. Two immature specimens, 3 mm to 5 mm along the tentacular axis, were provisionally assigned to this species (Matthews, 1954). They were found creeping over the red alga, *Hypnea nidifica,* in Kaneohe Bay, Oahu. The ground color of the specimens was yellowish green, slightly denser at the bases of the tentacular sheaths and along the main tentacular axis. High magnification revealed extremely minute, bright red spots forming a narrow, irregular band on the peripherial upper surface. The branched tentacles were two to three times longer than the body. Eight aboral papillae (dorsal tentacles) were arranged in two equal rows. *C. duboscqui* was previously reported from Indo-China (Vietnam) waters in association with a pennatulacean (sea pen) (Dawydoff, 1930, 1938).

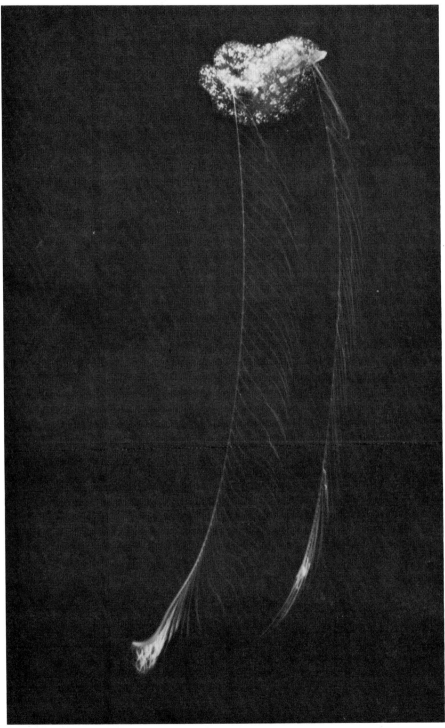

Figure 2.—CTENOPHORA. Coeloplanid collected from green alga *Halimeda* at Kahala, Oahu, in 1976; body length approximately 1 cm, tentacles about one-half extended (photo courtesy of Dale Sarver).

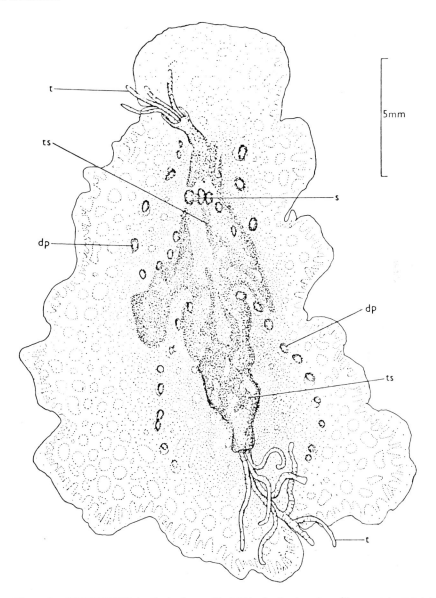

Figure 3.—CTENOPHORA. *Coeloplana willeyi* Abbott: *dp*, dorsal papillae; *s*, statocyst; *t*, tentacle; *ts*, tentacular sacs (from Gordon, 1969, *New Zealand J. Marine and Freshwater Res.* 3(3):469).

Coeloplana willeyi Abbott 1902. Hawaiian specimens were reported by Matthews and Townsley (1964) to vary in color from deep purple, red, or pink, to yellowish white depending on their location among the spines of the sea urchin *Echinothrix diadema*. Individuals range in length from 2 mm (contracted) to 32 mm (relaxed). The number of aboral papillae is variable (18 to 24) and difficult to determine, for they are not symmetrically arranged. According to Gordon (1969), the presence of white spots is an important character of *Coeloplana willeyi* (Fig. 3), yet these were

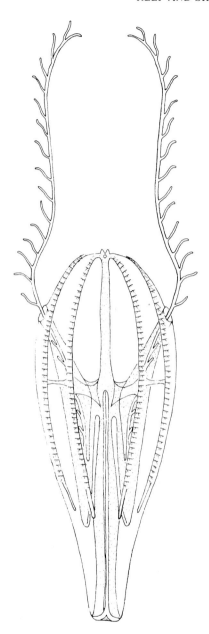

Figure 4.—CTENOPHORA. *Hormiphora palmata* Chun (redrawn from Mayer, 1906, by David Kemble, Bishop Museum).

Figure 5.—CTENOPHORA. Unidentified cydipped (from Edmondson, 1946).

said to be only "weakly developed" in Hawaiian specimens (Matthews and Townsley, 1964).

While *C. willeyi* is reported only from *E. diadema* in Hawaii this coeloplanid was found on "red *Zostera*" (Abbott, 1907), the alga *Sargassum* (Komai, 1922), and the holothurian, *Holothuria leucospilota* (Tokioka, 1969) in Japanese waters as well as on a bryozoan colony cast ashore on the northern coast of New Zealand (Gordon, 1969).[1]

Coeloplana bannwarthi Krumbach 1933. In a study on the symbionts associated with Indo West-Pacific diademid sea urchins, coeloplanids, dark red in color and about 2 cm long by 1 cm broad, were found wrapped around the large spines of *Echinothrix calamaris* and *Diadema paucispinum* in Hawaiian waters and these were determined as *C. bannwarthi* (R. U. Gooding, unpubl. record). This species was reported first from the spines of *Diadema* sea urchins in the Red Sea.

Coeloplana sp. Ten specimens of an undetermined coeloplanid, reaching a maximum length of 7 mm, have been collected from a sea star, *Acanthaster planci,* in Honolulu Harbor. When alive, their color was a very faint translucent yellow, with 15 to 20 diffuse but distinct reddish blotches on the upper side. No aboral papillae were evident on the living animals during the short period of observation. Specimens are deposited in Bishop Museum.

Order CYDIPPIDA

Hormiphora palmata Chun 1898. This species is spindle-shaped, about 40 mm long (Fig. 4), and has two long, branched tentacles arising from sheaths halfway down the length of the body. The rows of ctenes extend not quite two-thirds down the sides of the body. Each row is made up of about 40 combs. This species was first reported from surface waters around Oahu, Kauai, and Maui by Mayer (1906) as *Hormiphora fusiformis,* but Bigelow (1912) considered it as *H. palmata,* a species widespread in many other Pacific areas.

Two unidentified cydippids were reported by Edmondson (1933, 1946) from shallow waters surrounding Oahu. One spherical form, about 3 mm in diameter (Fig. 5), has comb rows extending the entire length of the body. It was observed

---

[1]The records of *C. willeyi* in association with the sea urchin *Heterocentrotus mammillatus* (Matthews and Townsley, 1964, p. 351; Fricke, 1970, p. 226) are inaccurate, based on the association of *C. weilli* Dawydoff 1938 with this sea urchin in the Gulf of Siam.

"occasionally" on the Waikiki reef. The other form, somewhat larger, has comb rows extending only about half the distance toward the mouth. It was reported as "common in the surface waters of Pearl Harbor." According to Matthews (1954) the above cydippids and other "undetermined" forms were commonly taken in Kaneohe Bay.

Order LOBATA

A representative of the order Lobata was reported from Hanauma Bay, Oahu (Matthews, 1954). It was noted to be small and colorless with large oral lobes and unsheathed tentacles.

Order CESTIDA

Specimens, possibly of a species of *Velamen,* with a compressed, ribbonlike shape, were sighted in Hanauma Bay and also reported by Matthews (1954).

GLOSSARY (CTENOPHORA)

aboral papillae: small raised extensions of the body on the upper (aboral, dorsal) side which may occur irregularly or in rows in some coeloplanids.
aboral sense organ: a sensory complex located opposite or away from the mouth and functioning in balancing the animal as well as in coordinating the beat of the comb rows.
auricles: flat, pointed flaps with ciliated edges formed by the two comb rows flanking the tentacles on each side in members of the order Lobata.
biradial symmetry: arrangement of the body in which the animal can be cut into identical, mirror-image halves in two planes; in ctenophores both such planes extend from the oral to the aboral surfaces, one running in the plane of the long axis of the mouth, the other at right angles to it.
colloblasts: adhesive epidermal cells on the tentacles used to capture prey.
ciliated bands: see **comb rows.**
ciliated plates: see **combs.**
comb plates: see **combs.**
comb rows: bands of combs extending along the body for some distance, in eight or fewer rows.
combs: or **ctenes,** specialized short transverse plates of elongate fused cilia, occurring as comb rows and providing the locomotor power by waves of ciliary movement.
ctenes: see **combs.**
dorsal papillae: see **aboral papillae.**
dorsal tentacles: see **aboral papillae.**
oral lobes: paired inflated expansions, one on each side of the mouth, of the flattened bodies of members of the order Lobata.
planktonic: small floating organisms carried by currents.
sense organ: see **aboral sense** organ.
statocyst: a pit or depression located within the aboral sense organ, and containing nerve and ciliary elements as well as calcareous grains; a balance organ.
symbiont: see symbiotic
symbiotic: an association between dissimilar organisms in which one serves as a host and others as associates (symbionts) living on the host.
tentacles: a pair of highly contractile filaments, often quite long, having short lateral branches in a single row and consisting of a solid muscular core covered with an outer layer containing colloblasts.
tentacular sacs: see **tentacular sheaths.**
tentacular sheaths: pouches containing the tentacles.

REFERENCES (CTENOPHORA)

Abbott, J. F.
    1907. The Morphology of Coeloplana. *Zoologische Jahrbücher Anatomie und Ontogenie* 24(1): 41–70. 7 figs, 3 pls.

Agassiz, A., and A. G. Mayer
    1899. Acalephs from the Fiji Islands. *Bull Mus. Comparative Zoology Harvard* 32(9): 157–189. 17 pls.
    1902. Medusae. Reports of the Scientific Results of the Expedition to the Tropical Pacific by the U. S. F. C. Steamer *Albatross* ... 1899–1900. *Mem. Mus. Comparative Zoology Harvard* 26(3): 137–176.

Bigelow, H. B.
    1912. The Ctenophores. Reports of the Scientific Results of the Expedition to the Eastern Tropical Pacific by the U.S.F.C. Steamer *Albatross* ... 1904–1905. *Bull. Mus. Comparative Zoology Harvard* 54(12): 369–404.

Dawydoff, C.
    1930. *Coeloplana duboscqui* nov. sp., Coeloplanide Provenant du Golfe de Siam, Commensale des Pennatules. *Arch. Zoologie Expérimentale et Générale* 70: 87–90.
    1938. Les Coeloplanides Indochinoises. *Arch. Zoologie Expérimentale et Générale* 80: 149–161. 16 figs., 1 pl.

Edmondson, C. H.
    1933 and 1946. *Reef and Shore Fauna of Hawaii.* B. P. Bishop Mus. Spec. Publ. 22. Honolulu.

Fricke, H. W.
    1970. Neue kriechende ctenophoren der Gattung *Coeloplana* aus Madagaskar. *Marine Biology* 5(3): 225–238. 16 figs.

Gordon, D. P.
    1969. A Platyctenean Ctenophore from New Zealand. *New Zealand J. Marine Freshwater Res.* 3(3): 466–471. 2 figs.

Komai, T.
    1922. *Studies on Two Aberrant Ctenophores,* Coeloplana *and* Gastrodes. Kyoto: Publ. by the author.

Matthews, D. C.
    1954. Records of Hawaiian Ctenophora. *Trans. American Microscopical Soc.* 73(3): 282–284.

Matthews, D. C., and S. J. Townsley
    1964. Additional Records of Hawaiian Platyctenea (Ctenophora). *Pacific Science* 18(3): 349–351.

Mayer, A. G.
    1906. Medusae of the Hawaiian Islands Collected by the Steamer "Albatross" in 1902. *Bull. United States Fish Commn.* (1903), pt. 3, pp. 1131–1143.

Tokioka, T.
    1969. A Creeping Ctenophore Found on the Sea Cucumber, *Holothuria leucospilota* (Brandt). *Publ. Seto Marine Biological Lab.* 17(4): 279–283. 1 fig.

INDEX*

A

Acabaria bicolor 125, **126, 127**
Acanthaster planci 159, 160, 267
Acontiferidae 145
Acropora 5, 178, 179, 180
　　echinata 163, 178, 180
　　paniculata 162, 179, **179**, 180
Acroporidae 178
Actiniaria 131–144
　　glossary 145
　　references 147
　　study techniques 131
Actiniidae 134
Actiniogeton 132
　　sesere 136, **136**
Adocia gellindra 67
Adociidae 65
Agariciidae 184
Aiptasia 132
　　californica 143
　　pulchella **142**, 143
Aiptasiidae 143
Alcyonacea 119–120, **121, 122, 123**
Alcyonaria 119
Alcyoniidae 119
algae 4, 6, 7, 8, 10, 60, 82, 138, 159, 160, 204, 262, 263, 264, 267
　　commensal 19, 120, 151, 152, 158–159, 190, 197
Aliciidae 134
Allogromiina 20
alpheid shrimps 4, 83, 224
Alpheus deuteropus 75
Alveolina melo 32
Alveolinella 36

Alveolinellidae 19, 21, 32, 34
Alveopora dedalea 163
　　verrilliana 163
Amansia 101
Amia (= Apogon) frenatus 114
Ammomassilina 28
　　alveoliniformis 26, **27**, 28
Ammonia 42
　　beccarii tepida 19, 20, 40, **41**, 42
Amphisorus 34
　　hemprichii 18, 30, **31**, 34
Amphistegina 42
　　cumingii 32
　　madagascariensis 16, 18, 20, 42, **43**
Amphisteginidae 19, 22, 42
Anemonia 132
　　mutabilis 134, **134**
Anisopsammia amphelioides 164
Anomalina 48
　　glabrata 46, **47**, 48
　　sp. 46, **47**, 48
Anomalinidae 22, 40, 46, 48
Anthelia edmondsoni 120, **123**
Anthemiphyllia pacifica 164
Antheopsis 132
　　papillosa **137**, 138
Anthopleura 132
　　nigrescens 135, **135**
　　sp. *a* 136
　　sp. *b* 136
Anthosigmella valentis 67
Anthothoe 132
　　sp. **141**, 142
Anthozoa 70, 119–129
Antipatharia 125, 242–261
　　glossary 255
　　references 261

Antipathes 243, 244, 247
　　dichotoma cover photo & legend, 243, 244, **252**, 253, **253**
　　galapagensis 253
　　grandis 242, 243, **243**, 244, 253, **254**
　　intermedia 244, 255, **259**
　　irregularis 243
　　japonica 255
　　lentipinna 253
　　punctata 244, 253, **256**
　　sp. 244, **251**, 253
　　subpinnata 244, **249**, 251
　　ulex 244, 255, **259**
　　undulata 244, **248**, 251
Aplysilla rosea 64
　　sulphurea 58, 64
　　violacea 56, 64
Aplysillidae 64
Apogon frenatus 114
Articulina 26
　　antillarum 28
　　pacifica 26, **27**
Ascobius simplex **14**, 15
Asteropus kaena 56, 66
Astraea (Orbicella) ocellina 163
Astraea rudis 163
Aurelia 108
　　cf. *labiata* 108, 111, **112**
Axechina lissa 67
Axinella solenoides 67
Axinellida 66, 67
Axociella kilauea 59, 65

B

Baeolidia major 133
Balanophyllia affinis 162, 165, 197
　　desmophyllioides 164
　　diomedeae 164

*Bold face page numbers indicate illustrations.

Balanophyllia affinis (cont.)
    hawaiiensis 162, 164, 165, **166**, 197, **198**
    laysanensis 164
    sp. 162, 165, **166**, 197, **198**
Balanophyllidae 197
bamboo coral 125
Bathyactis hawaiiensis 164
Beroe australis 262–263
    forskali 262, **263**
Beroida 262–263
Biloculina ringens var. *denticulata* 26
black coral 126, 158, 242–261, **cover photo & legend**
blue coral 119
Bolivina 39
    compacta 34, **35**, 39
    limbata 34
    striatula 19, 34, **35**, 39
Bolivinella 38
    folia 34, **35**, 38
Boloceroides 132
    mcmurrichi 132, **133**
Boloceroididae 132
Borelis 34
    melo 32, **33**, 34
Bougainvillia ramosa 84, **86**
Bougainvilliidae 72, 83
brain coral 180, 197
bryozoans 49, 267
Buliminella 39
    elegantissima 19
    milletti 34, **35**, 39
Buliminidae 21, 34, 36, 38, 39
Bunodactis manni 137
Bunodeopsis 132, 134
    medusoides 133, **133**

C

Cadonematidae 72
Calcarea 53, 64, 67
calcareous algae 6, 7, 8, 159
calcareous sponges 54
Calcarina 40, 44
    murrayi 40, **41**, 44
Calcarinidae 25, 40, 42
Calliactis 132
    armillatus 141
    polypus 141, **141**, 143
Callyspongia diffusa 64, 65
Callyspongiidae 65
Calyptoblastea 72, 87
Camerina 38
Camerinidae 19, 21, 32, 38
Campanulariidae 71, 72, 88
Campanulinidae 72, 90

Candeina 19, 48
    nitida 44, **45**, 48
carangid fish commensal with scyphozoans 114, 116
Caranx mate 114
Carmia contarenii 65
Caryophyllia alcocki 164
    hawaiiensis 164
    octopali 164
Cassidulina 46
    delicata 44, **45**, 46
    minuta 44, **45**, 46
Cassidulinidae 21, 44
Cassiopea 109, 112
    medusa 109, **112**, 113
    mertensi 109, 113
Cassiopeidae 112
Caulerpa serrulata 82
Cephea cephea 109, 113, **114**
Cepheidae 113
Ceriantharia 145
    glossary 145
    references 147
Ceratella fusca 74
Ceratotrochus laxus 164
Cerianthus sp. 145
Cestida 268
Charybdea alata 108, 110, **111**
    arborifera 110
    moseri 110
    rastoni 108, 110, **111**
Chironex fleckeri 109
Chondrophora 71, 105–107
Chondrosia chucalla 63, 66
Chondrosiidae 66
Choristida 66, 67
Chrysalidina dimorpha 36
Chrysalidinella 39
    dimorpha 36, **37**, 39
Cibicides 46, 48
    lobatulus 46, **47**, 48
    sp. 20, 40, **41**, 48
Ciliata 12–15
Ciliophora 12–15
Ciocalypta penicillus 65
Cirrhipathes 243, 244, 247
    anguina 244, **247**, 249
    spiralis 244, **246**, 249
Cladactella 132
    manni **136**, 137
    obscura 135
Cladonema 79
    radiatum 77, **78**
Cladonematidae 72, 77
Clathria procera 63, 65
Clathrina sp. 64, 67
Clavidae 72, 81
Cliona vastifica 58, 66

Clionidae 66
Clytia 89
    hemisphaerica 89, 90, **90**
Cnidaria 70–261
Cnidospora 12
Coelastrea tenuis 163, 203
Coelenterata 70–261, 262
Coeloplana 263
    bannwarthi 267
    duboscqui 263
    sp. 267
    willeyi 265, **265**, 267
coeloplanid **264**, 267
Coenothecalia 119
comb jellies 262
commensal algae 19, 120, 151, 152, 158–159, 190, 197
Coralliidae 125
Corallimorpharia 130
    glossary 145
    references 147
Corallimorphidae 130
Corallium 125
corals 2, 3, 5, 8
    bamboo 125
    black 126, 158, 242–261, **cover photo & legend**
    blue 119
    brain 180, 197
    finger 160, 181, 186, **220**, 221, 223
    fire 71
    gold 125
    hermatypic 158, 159, 160, 161
    hydrozoan 158
    mushroom 159, 204
    pink 158
    precious 125
    rose 217
    scleractinian 158
    soft 119, 120, 148, 158
    study techniques 165
Cordylophora caspia 81, **82**
    lacustris 82
Cornuspira 30, **49**
    planorbis 19
Corydendrium splendidum 71, 80, 81
Corynactis sp. 130, **130**
Coryne 75
Corynidae 72, 75
Coscinaraea 233
    ostreaeformis 162, 169, **170**, **232**, 233
Cotylorhizoides pacificus 114

INDEX

crustaceans 2, 3, 4, 6, 7, 8, 9, 75, 83, 105, 134, 141, 142, 212, 242, 244
Ctenophora 262-268
   glossary 268
   references 269
Cubomedusae 108, 109-111
Culcita novaegineae 197
Cuspidella costata 91
   sp. 91, **91**
Cyathoceras diomedeae 164
Cycloseris 208, 212
   fragilis 162, 163, **166**, 167, 209, **210**, 212
   hexagonalis 162, **166**, 167, **211**, 212
   patelliformis 163
   vaughani 162, 163, 165 **166**, 208, **208**, **209**
cydippid 267, **267**
Cydippida 267-268
Cymbalopora poeyi var. *bradyi* 42
Cymbaloporella 46, **49**
Cymbaloporetta 46, 48, **49**
   bradyi 42, **43**, 46
   squammosa 20, 42, **43**, 46
Cymbaloporidae 22, 42, 44
Cyphastrea 201
   ocellina 162, 163, 171, **173**, **200**, 201

D

Damiriana hawaiiana 60, 63, 65
damselfish 221
Dardanus 141, 142
Dascyllus 221
Deltocyathus andamanicus 164
Demospongiae 53, 54, 55, 64-67
Dendrilla cactus 58, 64
Dendroceratida 53, 64
Dendrophyllia manni 163, 197
   oahensis 164
   serpentina 164
Dendrophyllidae 197
Densa distincta 66
Desmacidonidae 65
Desmophyllum cristagalli 164
Diadema 267
   paucispinum 267
diademid sea urchins 267
Diadumene 132
   leucolena 143, **143**
Diadumenidae 143

Dictyoceratida 54, 64
Dictyosphaeria 60, 160
Diplastrella spiniglobata 66
Diplopilus couthouyi 113
Discorbidae 22, 36, 39, 40, 44
Discorbina patelliformis 40
Discorbis 40
   mirus 40, **41**
   orientalis 36
   patelliformis 40
   sp. 19
Donatia deformis 54
dörnchenkorallen 242
Dorypleres pleopora 66
Dynamena cornicina **94**, 95
   crisoides 93, **93**, 95, 96
Dysidea avara 64
   herbacea 58, 64
   sp. 58, 64
Dysideidae 64

E

echinoderms 2, 3, 4, 5, 6, 9, 10, 197, 262, 265, 267
Echinothrix calamaris 267
   diadema 265, 267
ecosystems 2-3, 4-10
Edwardiella carneola 140
Edwardsia 132
   sp. *a* **143**, 144
   sp. *b* 144, **144**
Edwardsidae 144
Elasmopus calliactis 142
Eleutheria 71, 79
   bilateralis 79
   oahuensis 79
Eleutheriidae 72, 78
Elphidium 20, 36
   advenum 34, **35**, 36
   gunteri galvestonensis 19
   poeyanum 34, **35**, 36
   sp. 34, **35**, 36
Endopachys oahense 164
Entosolenia marginata 34
Epiphellia 132
   humilis 140, **141**
   pusilla 140, **140**
Epitonium ulu 208
Erylus proximus 66
   rotundus 67
Eudendriidae 72, 86
Eudendrium 87
   capillare 87
   sp. 87, **87**
Eufolliculina lignicola 13, **14**, 15
Eurypon distincta 66, 67
   nigra 59, 66

F

faunal characteristics 1-2, 4-10, 18, 54, 71, 105, 108, 119, 131, 161
   references 10
Favia hawaiiensis 163, 203
   hombroni 163
   rudis 163
   speciosa 163
   stelligera 163
Faviidae 197
Fijiella simplex 36
finger coral 160, 181, 186, **220**, 221, 223
fire coral 71
fish 4, 5, 6, 7, 8, 9, 10, 114, 116, 221, 242
Fissurina 39
   marginata 34, **35**, 39
Flabellum deludens 164
   pavoninum 164
Flintina 30
   bradyana 26, **27**, 30
Florius 19
Folliculinidae 12
Foraminifera 8, 12, 15-49
   glossary 50
   references 51
   study techniques 16
Fungia 159
   dentigera 163
   echinata 163
   fragilis 163, 209
   oahuensis 163, 205
   patella 163, 208
   paumotensis 163
   (Pleuractis) scutaria 162, 163, 165, **166**, 167, 204, **206**, **207**, 208, 209
   scutaria 10
Fungiidae 204
Fusulinina 20

G

gall crab 212
Gardineria hawaiiensis 164
Gaudryina 22
Garveia sp. 84, **85**
Geodia gibberella 56, 66
Geodiidae 66
Gerardia 125
Geryoniidae 71
Glabratella patelliformis 40
Globigerina 19, 48
   conglobata 44
   eggeri 44, **45**, 48
   sacculifer 44
Globigerinidae 22, 44, 26, 48

Globigerinoides 19, 48
   conglobatus 44, **45**, 48
   sacculifer 44, **45**, 48
Globorotalia 19, 48
   menardi 44, **45**, 48
   sp. 44, **45**, 48
Globorotaliidae 22, 44, 48
glossaries
   Actiniaria 145
   Antipatharia 255
   Ceriantharia 145
   Corallimorpharia 145
   Ctenophora 268
   Hydroida 102
   Octocorallia 128
   other hydrozoans 107
   Porifera 68
   Protozoa 50
   Scleractinia 237
   Scyphozoa 117
   Zoanthiniaria 156
Gnathanodon speciosus 114
gold coral 125
Gorgonacea 124–127
gorgonians 242, 243
Grantiidae 67
Gymnoblastea 72, 73
Gypsina 48
   globula 46, **47**, 48

H

Hadromerida 66, 67
Halecium 87
   beani 87, **88**
Haleciidae 72, 87
Haliclona 53
   aquaeducta 62, 65
   flabellodigitata 67
   permollis 62, 65
Haliclonidae 65
Halichondria coerulea 62, 66
   dura 60, 66
   melanadocia 54, 62, 66
Halichondrida 65
Halichondridae 65
Halimeda 264
Halocordyle 80, 81, 83, 84, **85**, 86
   disticha 80, 81, **81**
Halocordylidae 72, 79
Halofolliculina annulata 13, **14**, 15
Halopteris diaphana 71, 99, **100**
Hapalocarcinus marsupialis 212
Haplosclerida 65, 67

Hauerina 28
   bradyi 26, **27**, 28
   involuta 26
   pacifica 20, 26, **27**, 28
Hawaiian marine fauna
   characteristics 1–2, 4–10, 18, 54, 71, 105, 108, 119, 131, 161
   references 10
hermatypic corals 158, 159, 160, 161
hermit crabs 141, 142
Herviella mietta 136
Heteranthus 132
   verruculatus 138, **139**
Heterocentrotus mammillatus 267
Heterohelicidae 21, 34, 38
Heterostegina 38
   suborbicularis 16, 18, 20, 32, **33**, 38
Heterotrichida 12–15
Hexacorallia 119
Hexactinellida 53
Hexadella pleochromata 64
Hiattrochrota protea 65
Hippa pacifica 8, 105
Holothuria leucospilota 267
Homaxinella anamesa 67
Homosclerophorida 67
Homotrema 49
Homotremidae 21, 46, 49
Hopkinsina pacifica 19
Hormiphora fusiformis 267
   palmata **266**, 267
Hormathiidae 141
Hydroida 71–104
   glossary 102
   references 103
Hydrozoa 70–107
   glossary 107
   references 71, 107
hydrozoan corals 158
Hymedesmia sp. 59, 65
Hymeniacidon chloris 62, 66
Hymeniacidonidae 66
Hypnea nidifica 263

I

introduction references 10
Iotrochota protea 63, 65
Isarachnanthus bandanensis **144**, 145
Isaurus 148
   elongatus 148, **149**, 151
Isididae 125
Isophellidae 139

J

Jaspidae 66
Jaspis pleopora 66
jellyfish 2, 108, 114, 116

K

Kaneohea poni 65
Kishinouyea hawaiiensis 108, 109, **110**
   pacifica 109
Koleolepas tinkeri 142
Kotimea tethya 67

L

Lagena 36, 39
   globosa 32, **33**, 36
   marginata 34
Lagenidae 21, 32, 36
Lagotia viridis **14**, 15
Leiodermatium sp. 56, 67
Leiopathes 244, 247
   glaberrima 244, **250**, 251
Leptastrea 202
   agassizi 163, 202
   bottae 162, 163, **173**, 174, **201**, **202**, **202**
   hawaiiensis 163, 202
   purpurea 162, 163, 174, **175**, 203, **203**, **204**, **205**
   stellulata 163
Leptoseris 184
   digitata 163, 190, **194**
   hawaiiensis 162, **168**, 169, 190, **195**
   incrustans 162, 167, **168**, 186, 190, **193**
   papyracea 162, 163, 167, **168**, 190, **194**
   scabra 162, **168**, 169, 190, **195**
   tubulifera 162, 167, **168**, 190, **196**
Leucetta solida 67
   sp. 64, 67
Leucettidae 67
Leuconia kaiana 64, 67
Leucosolenia eleanor 67
   vesicula 67
Leucosolenida 67
Leucosoleniidae 67
limu make o Hana 154
Liriope 71
   tetraphylla 71
Lissodendoryx calypta 67
Lithistida 67
little thorn corals 242
Lobactis danae 163
Lobata 268

INDEX

Loxostomum 39
    limbatum 20, 34, **35**, 39
Lybia edmondsoni 134, **134**
Lytocarpus philippinus 71

M

Macranthea cookei 138
Madracis kauaiensis 164
Madrepora kauaiensis 163, 164
Madreporaria 158
Marginopora 34
    vertebralis 16, 18, 20, 30, **31**, 34
Massilina 28
    agglutinans 22
    alveoliniformis 26
    crenata 22, **23**, 28
    secans **24**, 25, 28
Mastigias ocellatus 114
Mastigiidae 114
Melithaeidae 126
mermaid's penny 16
Metafolliculina andrewsi **14**, 15
    nordgardi 13, **14**, 15
Microciona haematodes 67
    maunaloa 60, 65
    sp. 59, 65
Microcionidae 65
Miliola trigonula 28
Miliolidae 21, 22, 25, 26, 28
Miliolina 21
    alveoliniformis 26
    oblonga 25, 28
    parkeri 25
    transversestriata 25
Millepora miniacea 46
Milleporina 71
Miniacina 49
    miniacea 46, **47**, 49
Mirofolliculina limnoriae **14**, 15
Moerisia 73
Moerisiidae 72, 73
mollusks 2, 3, 4, 5, 6, 7, 8, 9, 142, 143, 144, 159, 208, 242
Monalysidium 32
    politum 20, 32, **33**
Montipora 180, 230
    bernardi 163, 180
    capitata 163, 182
    dilatata 162, 171, **173**, 182, **183**
    flabellata 162, 163, 171, **172**, 182, 183, **184**
    incognita 163
    patula 162, 163, 171, **172**, 183, **185**
    studeri 163, 180

Montipora (cont.)
    tenuicaulis 163, 180
    venosa 163, 182
    verrilli 162, 171, **172**, 183, **187**
    verrucosa 9, 162, 163, 171, **172**, 180, **181**, **182**, 183
mushroom corals 159, 204
Mussa 163
Mycale cecilia 54, 60, 63, 65
    (*Carmia*) *contarenii* 62, 65
    manuakea 65
    maunakea 65
Mycalidae 65
Myriastra debilis 66
Myxilla rosacea 63, 65
Myxillidae 65

N

Naniupi ula 59, 65
Nautilus acicularis 32
    arietinus 32
    lobatulus 46
    melo 32
    pertusus 28
    planatus 28
Nectothela lilae 132
Neoadocia mokuoloe 65
Neoconorbina 40
    patelliformis 20, 40, **41**
Nodophthalmidium 30
    antillarum 28, **29**, 30
Nonion 38
    boueanum 34, **35**, 38
    sp. 19
Nonionella 38, **49**
Nonionidae 21, 34, 36
Nonionina boueana 34
Nummulites 38
    cumingii 32

O

Obelia australis 89
    dichotoma 89, **89**
Octocorallia 119–129
    glossary 128
    references 128
Ocypode ceratophthalmus 8, 105
    laevis 8, 105
Oolina globosa 32
Operculina 38
    philippinensis 20, 32, **33**, 38
Operculinella 38
    cumingii 32, **33**, 38
Ophthalmidiidae 21, 28, 30

Orbitolites marginalis 30
Orbulina 19, 46, 48
    universa 44, **45**, 48
Orthopyxis 89
Oscarella tenuis 67
Ostroumovia horii 73, 74
oysters 7, 144

P

Padina 7, 138
pagurid crabs 141, 142
Palythoa 9, 148, 151, 154
    psammophilia **151**, 152
    toxica 152, **152**, 156
    tuberculosa **150**, 151
    vestitus **150**, 151, 154
paper shells 16
Paracyathus gardineri 164
    mauiensis 164
    molokensis 164
    tenuicalyx 164
Parafolliculina violaceae 13, **14**, 15
Parantipathes 244, 247
    sp. 244, 255, **260**
Pavona 184
    clavus 163, 186
    duerdeni 162, 163, **168**, 169, 186, **189**, **191**
    explanulata 163, 186
    (*Polyastra*) 186
    (*Pseudocolumnastraea*) 184
    (*Pseudocolumnastraea*) *pollicata* 162, **166**, 167, 186, **192**, **193**
    repens 163
    varians 162, 163, 169, **170**, 186, **188**, **189**
Pavonina 39
    flabelliformis 36, **37**, 39
Pelagia notiluca 108
Pellina eusiphonia 62, 65
    sitiens 65
Peneroplidae 19, 21, 28, 30, 32
Peneroplis 32
    arietinus 32
    cylindraceus 32
    pertusus 28, **29**, 32
    planatus 28, **29**, 32
Pennaria 80
    tiarella 71, 80, 81
Pennatulacea 119, 263
Petrosia puna 55, 65
Phellia humilis 140
Phestella melanobranchia 197
Phorbasidae 65
Phycopsis aculeata 67

Phyllorhiza 114
  pacifica 114
  punctata 109, 114, **115**
Phymanthidae 138
Physalia 105, 107, 110
  physalis 105, **106**
  utriculus 105
pink coral 158
Placotrochus fuscus 164
Plakina monolopha 62, 67
Plakinidae 67
Plakortis simplex 60, 67
Planispirillina 39
  denticulogranulata 36, **37**, 39
Planorbulina 48, **49**
  echinata 40
Planorbulinidae 22, 46, 48
Platyctenea 263–265, 267
Platygyra 180
Pleraplysilla hyalina 58, 64
Plumularia (Halopteris) buski 71
  margaretta 100, **101**
  setacea 101, **102**
Plumulariidae 71, 73, 99
Pocillopora **175**, 179, 186, 212, **213**, 217, 219
  aspera 163
  brevicornis 163, 212
  cespitosa 163, 212
  damicornis 162, 163, **176, 177**, 178, 212, **213, 214**, 215, **215**, 217
  elegans 163
  elongata 163
  eydouxi 162, 163, 164, **177**, 178, 212, **219, 220**
  favosa 163
  frondosa 163
  informis 163
  ligulata 162, 163, **177**, 178, 217, **218**
  meandrina 9, 10, 163, 164, **177**, 212, 217, 219, 221
  meandrina var. nobilis 162, 178, **216**, 217
  modumanensis 163, 220
  molokensis 162, 164, **177**, 178, 217, **218**, 219
  nobilis 163
  plicata 163
  rugosa 164
  solida 164
  verrucosa 164
Pocilloporidae 212
Poecilosclerida 65, 67
polychaete worms 4, 8, 159

Polydectus cupulifera 139, **140**
Polystomella advena 34
  poeyana 34
Polytrema miniaceum 46
Porifera 53–69
  glossary 68
  references 69
  study techniques 53
Porites **175**, 179, 217, 221, 223, 224, 225, 227, 229, 230
  brighami 162, 174, **176, 228**, 229
  bulbosa 164
  compressa 10, 162, 164, 174, **175, 176**, 178, 181, 186, **220**, 221, **222**, 223, 224, 227
  compressus 164
  duerdeni 162, 174, **175**, 223, **223**
  evermanni 162, 174, **175**, 225, **226**
  lanuginosa 164
  lichen 162, 164, 171, 229
  lobata 9, 10, 75, 77, 83, 97, 162, 164, 174, **176**, 224, **224, 225**, 227, 229
  mordax 164
  pukoensis 162, 164, **176**, 178, 227, **228**, 229
  quelchi 164
  reticulosa 164
  schauinslandi 164
  studeri 162, 174, **176**, 227, **227**
  tenuis 164
Porites (Synaraea) 230
  convexa 162, 164, 171, **172, 173**, 229, **230**, 231, **231**
  hawaiiensis 164, **229**, 230
  irregularis 162, 171, **173**, 231, **232**, 233
Poritidae 221
Poroeponides 42
  cribrorepandus 40, **41**, 42
Porpita pacifica **106**, 107
Portuguese man-of-war 105, 106, 110
precious coral 125
Primnoidae 125
Prosthiostomum montiporae 181
Prosuberites oleteira 59, 66
Protozoa 12–52
  glossary 50
  references 51
  study techniques 16

Psammaplysilla purpurea 58, 64
Psammocora 233
  nierstraszi 162, 169, **170, 236**, 237
  stellata 233
  verrilli 162, 169, **170**, 235, **236**, 237
Psammocora (Stephanaria) 233
  brighami 164, 233, **234**, 235
  stellata 162, 164, 169, **170**, 233, **234**, 235, **235**
Pseudocryptochirus 212
Pseudohauerina 28
  involuta 26, **27**, 28
  orientalis 26
  sp. 26, **27**, 28
Pseudomassilina 28
  cf. *P. agglutinans* 22, **23**, 28
Pseudononion 38
  japonicum 34, **35**, 38
Pulvinulina menardii 44
Pyrgo 30
  denticulata 26, **27**, 30

Q

Quinqueloculina 26
  agglutinans **24**, 25, 26
  alveoliniformis 26
  baragwanathi **24**, 25, 26
  bicarinata 26, 28, **29**
  bosciana 19
  ferussacii **24**, 25, 26
  granulocostata **24**, 25, 26
  laevigata 19
  parkeri 20, **24**, 25, 26
  poeyana 19, **24**, 25, 26
  polygona **24**, 25, 26
  pseudoreticulata 28
  secans 25
  sp. **24**, 25, 26
  sulcata **24**, 25, 26

R

Radianthus papillosa 138
Raspailiidae 66
Rectobolivina raphana 34
references
  Actiniaria 147
  Antipatharia 261
  Ceriantharia 147
  Corallimorpharia 147
  Ctenophora 269
  ecosystems 10
  faunal characteristics 10

INDEX

references (cont.)
    Hydroida 103
    Hydrozoa 71
    Octocorallia 128
    introduction to revised ed. 10
    other hydrozoans 107
    Porifera 69
    Protozoa 51
    Scleractinia 238
    Scyphozoa 118
    Zoanthiniaria 157
Reniera aquaeducta 65
Reussella 20, 38
    cf. *R. aequa* 36, **37**, 39
    simplex 36, **37**, 39
    spinulosa 36, **37**, 39
Rhaphisia myxa 60, 66
Rhizogeton sp. 81, 83, **84**
Rhizopodea 15–49
Rhizostomeae 109, 112–117
Rosalina 20, 40
    bulloides 42
    orientalis 36, **37**, 40
    squammosa 42
    cf. *R. vilardeboana* 36, **37**, 40
rose coral 217
Rotalia 44
    beccarii tepida 40
    menardii 44
    murrayi 40
Rotaliidae 25, 40
Rotaliina 21

S

Sagartia longa 139
    pugnax 134, 139
    pusilla 140
Sagartiidae 142
Sagrina raphanus 34
Sarcomastigophora 12, 15–49
Sarcothelia edmondsoni 120
Sargassum 5, 6, 267
Sarsia 75
    (Syncoryne) mirabilis 75, **77**
Schlumbergerina 30
    alveoliniformis 26, **27**, 30
Schizopathes 244, 247
    conferta 244, 255, **257**, **258**
scleractinian corals 158–241
    glossary 237
    references 238
    study techniques 165
Scyphozoa 70, 108–118
    glossary 117
    references 118

sea anemones 131–144, 145
sea fans 126, 158
sea pens 119, 263
sea urchin 267
sea wasps 109
Semaeostomeae 108, 111–112
sequin shell 16
Seriatopora 212
Sertularella diaphana 97
    speciosa 93, **95**, 96
Sertularia distans gracilis 93, **96**, 97
    ligulata 93, 97, **98**
    subtilis 93, 97, **99**
Sertulariidae 71, 72, 92
shrimps 2, 4
Siderastreidae 233
Sigmoilina 26
    costata 20, 22, **23**, 26
Sinularia abrupta 119, **120**, **121**, **122**
Siphogenerina 39
    raphana 34, **35**, 39
Siphonina echinata 40
Siphoninoides 20, 42
    echinatus 40, **41**, 42
Siphonophora 71, 105–107
soft corals 119, 148, 158
Solanderia fusca 74
    minima 73, 75
    misakinensis 75, **75**, **76**
    secunda 74
Solanderiidae 72, 73
Sorites 34
    marginalis 18, 30, **31**, 34
spicules of sponges 56, **57**, 60, **61**
Spirastrella coccinea 55, 60, 66
    keaukaha 66
    vagabunda 66
    valentis 67
Spirastrellidae 66
Spirillina 39, **49**
    denticulo-granulata 36
Spirillinidae 21, 36, **37**, 39
Spirolina 32
    acicularis 32, **33**
    arietina 20, 32, **33**
Spiroloculina 28
    angulata 20, 22, **23**, 28
    communis 22, **23**, 28
    corrugata 22, **23**, 28
    crenata 22
sponges 2, 53–69, 159
sponge spicules 56, **57**, 60, **61**
Spongia oceania 58, 64
Spongiidae 64
Sporadotrema 49

Sporozoa 12
Staurocladia 79
    acuminata 79
    alternata 79
    bilateralis 79, **79**
    oahuensis 79, **80**
Stauromedusae 108, 109, 110
Stelletiidae 66
Stelletta debilis 56, 66
Stellettinopsis kaena 66
Stephanophyllia formosissima 164
Stichopathes 243, 244, 247
    echinulata 244, **245**, 247
    paucispina 247
    robusta 247
Stoichactinidae 134
Stoichactis 132
    sp. **137**, 138
Stolonifera 119
stony corals 158
Streblus beccarii tepida 40
study techniques
    Actiniaria 131
    Protozoa 16
    scleractinian corals 165
    sponges 53
    Zoanthiniaria 148
Stylasterina 71
Stylochoplana inquilina 142
Stylophora 212
Suberitidae 66
Syntheciidae 73, 91
Synthecium tubitheca 91, **92**

T

Tealiopsis nigrescens 135
Tedania ignis 54, 59, 63, 65
    macrodactyla 60, 63, 65
Tedaniidae 65
Telestacea 120, 122, 124
Telestidae 120
Telesto 122
    riisei 122, **124**, **126**
Telmatactis 132
    decora 139, **139**, **140**
Terpios granulosa 59, 66
    zeteki 64, 66
Tethya 53
    diploderma 66
    cf. *diploderma* 54, 55, 66
    sp. 55, 66
Tethyidae 66
Textularia 25
    agglutinans 22, **23**, 25
    folia 34
    foliacea oceanica 20, 22, **23**, 25

Textularia (cont.)
    siphonifera 22, **23**, 25
Textulariidae 21, 22, 25
Textulariina 20
Thamastriidae 233
Thysanostoma flagellatum
    109, 116, **116, 117**
Thysanostomatidae 116
Timea xena 67
Tonna 142
Toxadocia violacea 59, 63, 65
Trapezia 212
Tretomphalus 19, 44
    bulloides 42, **43**, 44
Triactis 132
    producta 134, **134**
Trifarina 39
    bradyi 34, **35**, 39
    sp. 34, **35**, 39
Triloculina 30
    bicarinata 20
    cf. *T. bicarinata* 28, **29**, 30
    cf. *T. eburnea* 28, **29**, 30
    fichteliana 20, 28, **29**, 30
    linneana 28, **29**, 30
    oblonga 28, **29**, 30
    cf. *T. oblonga* **24**, 25, 30
    transversestriata **24**, 25, 30
    trigonula 28, **29**, 30
Trimosina simplex 36
Trochammina globigeriniformis 22, **23**, 25

Trochamminidae 21, 22, 25
Trochocyathus oahensis 164
Troglocarcinus 204
    crescentus 186
    minutus 202
Truncatulina lobatula 46
Tubastraea 197
    aurea 164, 197
    coccinea 162, 163, 164, 171, **173**, 197, **199, 200**
Turbo 143
Turritopsis nutricula 82, **83**

U

Ulosa rhoda 65
Ulva 5, 6, 78, 79
Uvigerina 39
    porrecta 34, **35**, 39
    raphanus 34

V

Vaginulinopsis 36
    tasmanica 32, **33**, 36
Velamen 268
Velella 107
    pacifica 107
    velella **106**, 107
Vermiculum globosum 32
    marginata 34
    oblongum 25, 28
Verneuilina spinulosa 36
Verongiidae 64

Vertebralina 30
    striata 28, **29**, 30

W

worms 2, 3, 4, 7, 8, 159, 242

X

xanthid crabs 208
Xeniidae 120
Xytopsiphum kaneohe 62, 65
    meganese 62, 65
Xytopsues zukerani 65

Z

Zaplethea digonoxea 55, 66
Zoantharia 119, 125, 130–157
Zoanthidae 148
Zoanthidea 148
Zoanthiniaria 148–157
    glossary 156
    references 157
    study techniques 148
Zoanthus 9, 148, 154
    confertus 154
    kealakekuaensis **155**, 156
    nitidus 154
    pacificus 154, **154**
    vestitus 151
zooxanthellae 19, 120, 151, 152, 158–159, 190, 197
Zostera 267
Zygomycale parishi 54, 58, 65